AF767436

Satellites Missions and Technologies for Geosciences

Edited by Vladislav Demyanov and Jonathan Becedas

Published in London, United Kingdom

IntechOpen

Supporting open minds since 2005

Satellites Missions and Technologies for Geosciences
http://dx.doi.org/10.5772/intechopen.83246
Edited by Vladislav Demyanov and Jonathan Becedas

Contributors
Vladislav Demyanov, Anna Yasyukevich, Maria Sergeeva, Daniel Garcia-Almiñana, Silvia Rodriguez-Donaire, Miquel Sureda, Santiago Perez, Eloi Sierra Salvadó, Peter Roberts, Jonathan Becedas, Georg Herdrich, Dhiren Kataria, Leonardo Ghizoni, Rachel Villain, Barbara Heißerer, Ameli Schwalber, Jonas Morsbøl, Kristian Bay, Victor Jungnell, Constantin Traub, Stefanos Fasoulas, Adam Boxberger, Francesco Romano, Nicholas Crisp, Vitor Oiko, Alexis Conte, Badia Belkouchi, Kate Smith, Steve Edmondson, Sarah Haigh, Rachel Lyons, Stephen Worrall, Sabrina Livadiotti, Claire Huyton, Luciana Sinpetru, Rosa María Domínguez, David González, Yung-An Chan, Ron Outlaw, Adriano Camps, Boris Gavrilov, Yuriy Poklad, Ilya Ryakhovskiy, Victor Nwankwo, Nnamdi Jibiri, Michael Kio, Shishir Priyadarshi, Giorgio Savastano, Michela Ravanelli, Sashibhushana Rao Gottapu, Bharati Bidikar, Babji Prasad Chapa, Mogadala Vinod Kumar, Vaios Lappas, Vassilis Kostopoulos

© The Editor(s) and the Author(s) 2020
The rights of the editor(s) and the author(s) have been asserted in accordance with the Copyright, Designs and Patents Act 1988. All rights to the book as a whole are reserved by INTECHOPEN LIMITED. The book as a whole (compilation) cannot be reproduced, distributed or used for commercial or non-commercial purposes without INTECHOPEN LIMITED's written permission. Enquiries concerning the use of the book should be directed to INTECHOPEN LIMITED rights and permissions department (permissions@intechopen.com).
Violations are liable to prosecution under the governing Copyright Law.

CC BY

Individual chapters of this publication are distributed under the terms of the Creative Commons Attribution 3.0 Unported License which permits commercial use, distribution and reproduction of the individual chapters, provided the original author(s) and source publication are appropriately acknowledged. If so indicated, certain images may not be included under the Creative Commons license. In such cases users will need to obtain permission from the license holder to reproduce the material. More details and guidelines concerning content reuse and adaptation can be found at http://www.intechopen.com/copyright-policy.html.

Notice
Statements and opinions expressed in the chapters are these of the individual contributors and not necessarily those of the editors or publisher. No responsibility is accepted for the accuracy of information contained in the published chapters. The publisher assumes no responsibility for any damage or injury to persons or property arising out of the use of any materials, instructions, methods or ideas contained in the book.

First published in London, United Kingdom, 2020 by IntechOpen
IntechOpen is the global imprint of INTECHOPEN LIMITED, registered in England and Wales, registration number: 11086078, 7th floor, 10 Lower Thames Street, London, EC3R 6AF, United Kingdom
Printed in Croatia

British Library Cataloguing-in-Publication Data
A catalogue record for this book is available from the British Library

Additional hard and PDF copies can be obtained from orders@intechopen.com

Satellites Missions and Technologies for Geosciences
Edited by Vladislav Demyanov and Jonathan Becedas
p. cm.
Print ISBN 978-1-78985-995-9
Online ISBN 978-1-78985-996-6
eBook (PDF) ISBN 978-1-78985-301-8

For EU product safety concerns:
IN TECH d.o.o., Prolaz Marije Krucifikse Kozulić 3, 51000 Rijeka, Croatia, info@intechopen.com or visit our website at intechopen.com.

We are IntechOpen, the world's leading publisher of Open Access books

Built by scientists, for scientists

4,900+
Open access books available

124,000+
International authors and editors

140M+
Downloads

151
Countries delivered to

Our authors are among the
Top 1%
most cited scientists

12.2%
Contributors from top 500 universities

WEB OF SCIENCE™

Selection of our books indexed in the Book Citation Index
in Web of Science™ Core Collection (BKCI)

Interested in publishing with us?
Contact book.department@intechopen.com

Numbers displayed above are based on latest data collected.
For more information visit www.intechopen.com

Meet the editors

Professor Vladislav Demyanov graduated from Irkutsk Military Aviation Engineering Institute as an electronic engineer in 1993, received his PhD in radio-wave physics from Irkutsk State University in 2000 and his doctorate of science degree in Engineering from Siberian Federal University (Krasnoyarsk, Russia) in 2011. He has been working for the Institute of Solar and Terrestrial Physics and for Irkutsk State Transport University as a senior researcher and as a Professor. As an expert in his field, he is a member of the Radio-engineering Science Council of Siberian Federal University and a Head of State Examination Commission of Moscow State Aviation University. His research interests are: ionosphere scintillation research, science education, navigation solutions for transport, and space weather impacts on GNSS.

Dr. Jonathan Becedas is an Electrical Engineer and Industrial Engineer who graduated with honors. He holds a Master in Research and a PhD in Mechatronics from the University of Castilla-La Mancha (UCLM), Spain. He is the author of nearly 100 fully refereed scientific papers and has participated in over 20 competitive research projects, many of them for NASA, ESA, and the European Commission. He held a position of Researcher and Associate Professor at the University of Castilla-La Mancha (Spain) from 2002 until 2009 and was a Research Associate at the University of Leicester (United Kingdom) from 2009 until 2011. He was Head of the Robotics Department in Ixion Industry and Aerospace from 2011 until 2013. In that year he joined Elecnor Deimos Satellite Systems as Research Manager. He is currently R&D Manager and Head of the Projects Department in Elecnor Deimos Satellite Systems, where he coordinates R&D projects and satellite missions.

Contents

Preface

Being a vital modern technology, satellite systems for navigation, telecommunication, and geosciences have developed rapidly in the last 25 years. In fact, we are on the verge of a massive explosion of new markets for GNSS and associated position determination applications. Plans for GNSS modernization are well advanced and will enable unprecedented positioning accuracy, signals availability, and greater system integrity over the globe. Currently, GNSS technologies have been accepted worldwide as an effective, accurate, and free tool for the geosciences. Numerous research projects have been realized in such vital areas and forms as GNSS-remote sensing (ionosphere and upper atmosphere), GNSS-Reflectometry (GNSS-R) and Satellite Altimetry (SA) (the soil moisture distribution, geodynamics and oceanic surface characteristics), and GNSS-Radio Occultation (GNSS-RO) for the atmospheric sciences.

Considering modern satellite technologies, we face the situation of "head and tail of the one medal". On the one hand the Earth ionosphere, multipath phenomena, and space weather impacts (such as solar radio flares and solar irradiation fluxes) have been the main challenge for all modern satellite-based technologies including GNSS and Low-Earth Orbit (LEO) constellations. It is well known that rapid fluctuations of a signal phase and amplitude fading (ionosphere scintillations), a signal envelope distortion due to multipath as well as powerful solar irradiation impacts, degrade the satellite systems performance. To achieve the highest quality of a satellite system we need to focus on the system robustness regarding the above-mentioned impacts. Modeling and theoretical analysis of scintillations appearance, multipath sources, and solar irradiation impact are the base of every solution to improve the performance of the satellite system for both practice and geosciences. On the other hand the satellite system reaction as a result of the scintillation, multipath, or solar irradiation impacts, may be considered to be very informative and promising way to obtain valuable information for research in geosciences.

And finally, we should bear in mind that all the GNSS-R, GNSS-RO, and SA research technologies are realized based on not only GNSS constellations but also on the variety of LEO satellites. The main value of these technologies is the great opportunities to obtain the characteristics of the Earth atmosphere, ionosphere, and Earth surface over the oceans and long-distant unpopulated territories with high accuracy and resolution. Modern advances in microelectronics, materials, combined with affordable and frequent launch opportunities, have led to a revolution which consists of small LEO satellite missions used for Earth observation and Space exploration. Small satellites are now being developed in large volumes for mega-constellations for Earth observation and low latency communications, thus democratizing space and making new space applications for both practice and geoscience a reality.

In summary, we can say that modern satellite technologies have become a base of our civilization and support our day-to-day activity in both practice and geosciences. This book is devoted to GNSS-remote sensing for ionosphere research, modeling and mitigation techniques to diminish the ionosphere and multipath impacts on GNSS, and survey of the modern satellite missions and technologies. We hope that the experts' opinions presented in the book will be interesting for the research community and students in the area of satellites and space missions as well as in engineering and geoscience research.

Vladislav Demyanov
Irkutsk State Transport University,
Russia

Section 1

GNSS-Remote Sensing for Ionosphere Research

Chapter 1

GNSS High-Rate Data and the Efficiency of Ionospheric Scintillation Indices

Vladislav V. Demyanov, Maria A. Sergeeva and Anna S. Yasyukevich

Abstract

The work discusses the efficiency of different ionospheric scintillation indices. The new index D2fi based on the GNSS carrier phase observable was introduced. We analyze the accuracy of the phase measurements, in particular its dependence on the GNSS equipment thermal noises, multipath and external noises, and presettings of Phase Lock Loop and Code Delay Discriminator. The performance of DROTI, *S4*, $\sigma\varphi$, and D2fi was considered for the case of high-rate data. The "sensitivity" and reliability of each index differs significantly and depends on the time resolution of the carrier phase data. The new index D2fi advantages are that it is easily derived and has a clear dependence on GNSS hardware and software features. D2fi was proven to be a useful tool to detect the small-scale ionospheric disturbances based on high-rate GPS carrier phase measurements.

Keywords: GPS, ionospheric scintillation indices, high-rate GNSS data

1. Introduction

GNSS data with high-rate sampling becomes more and more available worldwide [1, 2]. It provides opportunities for the better results in the field of ionospheric scintillation research. Standard ionospheric indices and parameters *S4*, $\sigma\varphi$, and *ROTI*-based indices and ionospheric total electron (TEC) are widely used for the ionospheric research as reliable and informative tools [3–6]. Unfortunately, their accuracy, efficiency, and reliability depend on the integration time, input data sampling rate, and de-trending and filtering procedures of the carrier phase time series [7–9]. GNSS hardware architecture, Code Delay Discriminator (CDD), and Phase Lock Loop (PLL) presets play the crucial role in the carrier phase measurement quality especially under multipath environment conditions [10, 11]. The mentioned issues bring uncertainties to the ionospheric indices calculations which, in turn, can degrade the experimental results interpretation. The sensitivity of the ionospheric indices/parameters depends on the time resolution of input data. One of the important questions is whether the data rate is high enough to be sure that all the necessary ionospheric information is derived. There are different works on GPS scintillation, for instance, [1, 7, 12], but still there is a room for the more profound analysis of the data of higher time resolution than 10 Hz. Such a high-rate data is often considered as a noise but it is not exactly the truth. The excellent results by [1] based on the amplitude and phase measurements with the data rate of 100 Hz demonstrated the new opportunity to look at, and far beyond, 10 Hz resolution. The ionospheric scintillations show different features at different GNSS frequencies. Hence, the single-frequency carrier phase measurements can be involved for more informative analysis. The final accuracy of the carrier phase measurements depends on the GNSS equipment internal noises, multipath, and external noises. Incorrect presettings of PLL and CDD as well as the bad quality of reference oscillator can mislead a researcher in his or her final conclusions. To mitigate the impact of the mentioned factors, it is necessary to preset the receiver hardware (including antenna, preamplifier, and inter-frequency filter) and software (PLL and CDD types and parameters).

It is important to find such an ionospheric scintillation index which is easily derived and has a clear dependence on both the ionospheric turbulence structure and GNSS hardware and software presets. In this work, the second-order derivative of the GPS signal carrier phase based on high-rate carrier phase time series is suggested as a promising means for the ionospheric scintillation detection. No additional complex processing is needed to obtain this new scintillation index.

The work [1] and the general necessity to define the GPS data time resolution sufficient for the robust scintillation analysis were the motivation for the authors to test the real sensitivity of the ionospheric indices depending on the input data sampling rate. We consider GNSS carrier phase observable to be the most capable of observing the ionospheric disturbances and scintillations. The aims of this study include (a) introduction of the new index that is the second-order derivative of the GPS signal carrier phase (D2fi index) which helps to reveal scintillation events; (b) test of sensitivity of D2fi, DROTI, S4, and $\sigma\varphi$ indices based on 50 Hz GPS data; and (c) consideration of the benefits and limitations of these indices for scintillation studies. The analysis was performed for the case study and was based on GPS data of the mid-latitude GNSS station located near Irkutsk, Russia, during the intense geomagnetic storm.

2. The carrier phase noise content at the phase lock loop input

Ionospheric phase scintillations are induced with ionospheric irregularities of hundreds of meters to several kilometer size. These irregularities correspond to the Fresnel frequencies from ≈ 0.1 to ≈ 10 Hz [13, 14]. According to [1, 2, 15], it is possible to detect small-scale ionospheric irregularities of hundreds of meters to several kilometer size by observing not only the fast carrier phase variations but also the carrier phase noise variations which were considered earlier as "noise" [1]. This is possible if the data sampling rate is high enough to exclude low-frequency variations and trends from the carrier phase time series. The data sampling rate should contain the sufficient ionospheric information. The authors [1] showed that the majority of the phase scintillation events can be revealed if data sampling rate between 10 and 40 Hz is used. Therefore, for the analysis of weak ionospheric scintillations, the sampling data rate higher than 10 Hz should be used.

To extract the phase noise variations from the complex carrier phase data, we use the carrier phase derivatives. The second-order derivative works as a high-pass filter and removes the phase ambiguity, all the low-frequency trends (due to the relative motion between satellite and receivers), multipath slow variations, and low-frequency phase variations due to reference oscillator frequency drift. It allows us to extract the phase noise variations from the phase measurements without additional complex processing procedures. The carrier phase noise derivative can be also used as a new parameter in GPS occultation technology [16].

Let us estimate the values of the main components of the carrier phase noise. They should be small enough to obtain the pure ionospheric phase scintillation based on the D2fi index. In case of the stationary receiver, there are no phase variations and phase measurement noises due to vibration and jerks. Based on this assumption, the noise error in carrier phase measurements depends on two main factors: the carrier-to-noise ratio at the PLL input and the multipath noises at the reception point.

For an ideal PLL without inner loss, the noise dispersion of phase measurements is determined as follows [17]:

$$\sigma_{\varphi}^{2}=\frac{1}{(2\pi)^{2}}\bullet\frac{\Delta F_{PLL}}{2\bullet CN_{0}} \tag{1}$$

where ΔF_{PLL} is the noise bandwidth of the PLL filter (Hz) and CN_0 is the carrier-to-noise ratio at the PLL unit input (dBW).

Thus, the noise level of the carrier phase measurements is determined by the carrier-to-noise ratio (CNR) at the PLL input. The CNR depends on (1) the level of external noises, (2) the antenna pattern, and (3) the low-noise preamplifier (LPA) gain. In addition to external noise, the inherit receiver thermal noise, the short-term instability of the reference oscillator, the signal sampling, and quantization noise should be considered as well.

According to expression (1), the final accuracy of the carrier phase measurements depends on the filter noise bandwidth. At the same time, the carrier-to-noise ratio at the PLL input depends on the time of accumulation of instantaneous phase measurement samples. Thus, the noise dispersion of phase measurements can be determined more precisely as follows [17]:

$$\sigma_{\varphi}=\frac{1}{(2\pi)^{2}}\bullet\sqrt{\sigma_{T}^{2}+\sigma_{F}^{2}} \tag{2}$$

where σ_T^2 is the dispersion of receiver thermal noise and σ_F^2 is the dispersion of noise caused by the Allan deviation.

The noise components of the phase measurements with dispersions σ_T^2 and σ_F^2 depend on the above factors as follows [17]:

$$\sigma_{T}=\frac{1}{2\pi}\bullet\sqrt{\frac{\Delta F_{PLL}}{CN_{0}}\bullet\left(1+\frac{1}{2T_{COR}\bullet CN_{0}}\right)} \tag{3}$$

$$\sigma_{F}=m\bullet\frac{\sigma_{F}(\tau)\bullet f}{\Delta F_{PLL}} \tag{4}$$

where T_{COR} is the time of accumulation of instantaneous phase measurement samples (ms), $\sigma_F(\tau)$ is the RMS of the short-term instability of the reference generator frequency (Hz), f is the signal carrier frequency, and m is a coefficient depending on the PLL filter type (m = 144 for a second-order PLL filter and m = 160 for a third-order PLL filter).

The carrier-to-noise ratio at the PLL input is a function of the receiver noise temperature (including the antenna), as well as the environment noise temperature (the Earth noise, the total noise of cosmic radio sources, and the Sun noise). The measurements of noise caused by analog-to-digital signal conversion, as well as signal-to-noise level with regard to filtering, amplification, and antenna gain, can be expressed through the corresponding loss in the resulting carrier-to-noise ratio at the phase detector input. Therefore, the carrier-to-noise ratio at the PLL input can be expressed as follows [17]:

$$CN_0 = P_{rec} + G_A - N_T - L_{tr} - L_{dg} \tag{5}$$

where *Prec* is the signal level at the receiving point (dBW); G_A is the antenna gain (dB); N_T is the spectral density of the receiver thermal noise power (dBW); L_{tr} is the total power loss during filtering, frequency conversion, and the signal attenuation in the cable (dB); and L_{dg} is the signal power loss due to its analog-to-digital conversion (dB).

According to formula (5), two factors affecting the carrier-to-noise ratio at the PLL input can be deduced. The first factor is constant during the measurement and depends on the receiver equipment type. It is defined by the L_{tr}, L_{dg}, and G_A values. These typical values are $L_{tr} = -2... -4$ dB, $L_{dg} = -0.55 ... -3.0$ dB, and $G_A = -2 ... -7.5$ dB (depending on the satellite line-of-site angular direction) [18].

At the same time, there is a factor that depends not only on the equipment type but also changes randomly. This is the receiver thermal noise N_T. Let us estimate its change limits and reveal the most significant causes that affect the magnitude of this noise. The spectral power of the thermal noise is related to the temperature of medium [18]:

$$N_T = 10 \bullet lg(K \bullet T_\Sigma) \tag{6}$$

where $K = 1.38 \bullet 10^{-23}$ (W•s/K) is the Boltzmann constant and T_Σ is the total noise temperature of the equipment and the external environment, forming measurement noise.

The total noise temperature can be estimated as follows [18]:

$$T_\Sigma = T_{EXN} + T_A + T_{LPA} \tag{7}$$

where T_{EXN} is external noise due to the Earth noise (T_{EN}), the noise of the Galaxy and cosmic radio sources (T_{CN}), and the Sun noise (T_{SN}); T_A is the antenna noise temperature caused by the active loss resistance noise in the antenna; and T_{LPA} is the noise temperature of a low-noise preamplifier.

Under standard physical conditions, the Earth noise temperature is $T_{\Sigma,EN} = 300K$. The Earth noise component, which is present at the PLL input ($T_{\Sigma,EN}$), is determined by the antenna pattern as follows [18]:

$$T_{EN} = 100 \bullet \left(\frac{\beta}{2\theta}\right)^2 \bullet T_{\Sigma,EN} \tag{8}$$

where $\frac{\beta}{2\theta}$ is the ratio of the angular aperture of a groundward part of the antenna pattern, with respect to the total angular aperture of the antenna pattern.

According to Eq. (8), the higher the $\frac{\beta}{2\theta}$ ratio, the higher the magnitude of the Earth noise. With regard to the known antenna pattern of typical navigation receiver antennas, the value $\left(\frac{\beta}{2\theta}\right)^2$ can be within 0.004–0.01 [18]. Thus, the Earth noise temperature at the PLL input is $T_{EN} = 1.2–3\ K$, and the correspondent noise spectral power varies from −227.8 to −223.8 dBW/Hz.

Similarly, the Sun noise temperature can be obtained. The total noise temperature of the Sun is $T_{\Sigma,S} = 6000K$. The angular size of the Sun visible from the Earth's surface is $\beta_C = 0{,}5°$. Considering the above mentioned typical antenna pattern, the $\left(\frac{\beta}{2\theta}\right)^2$ ratio is about 10^{-5}. When the sunlight falls into the antenna aperture, the Sun noise temperature $T_{SN} = 0{,}00001 \times 6000 \approx 0{,}06$ K. This corresponds to the Sun noise temperature at the PLL input of about -241 dBW/Hz.

The sky noise temperature (T_{CN}), including all cosmic radio noise sources, can be considered equal to 100 K [18]. This noise is accepted for the entire antenna aperture. If we consider an ideal antenna without losses, the corresponding sky noise power at the PLL input is about −208 dBW/Hz.

The inherit antenna noise temperature T_A results from the noise of active loss resistance in the antenna [18]:

$$T_A = T_0 \bullet (1 - \eta) \tag{9}$$

where T_0 is the antenna physical temperature and η is the antenna efficiency.

If the antenna temperature is equal to 300 K and the typical antenna efficiency is between 80 and 90%, the temperature $T_A \approx 60 \ldots 30$ K and the corresponding antenna noise power is between −211 and − 214 dBW/Hz.

The noise temperature of the preamplifier is defined as follows [18]:

$$T_{LPA} = T_{01} \bullet (\varepsilon - 1) \tag{10}$$

where ε is the preamplifier noise coefficient and T_{01} is the receiver physical temperature. Let us assume that the typical noise coefficient for the modern preamplifiers varies from 1.4 to 2 dB and T_{01} is 300 K. These conditions result in $T_{LNA} \approx 120 \ldots 300$ K, and the corresponding thermal noise power is from −204 to −208 dBW/Hz.

Table 1 shows the values of noise temperatures and noise spectral power for the above mentioned components of the receiver thermal noise (N_T) and external noises. According to the table, the thermal noise at the PLL input significantly depends on the receiver hardware and the antenna pattern. This can result in the significantly different carrier phase measurements accuracy and noise level when using navigation receivers and antennas of various types and models.

Using the information from the **Table 1** and formulas (2)–(5), we can estimate the noise level of the phase measurements in a stationary receiver when measuring the phase at different GPS frequencies and satellite elevations. Let us assume that $\Delta F_{PLL} = 18Hz$, accumulation time $T_{COR} = 20mc$, Allan deviation of the reference generator $\sigma_F(\tau) = 10^{-11}$, the maximum and minimum power levels of the signals (P_{rec}), received at L1, L2 and L5 frequencies are described by curves in **Figure 1** [19, 20]. The values of the standard deviation of the phase noise for this case are given in **Table 2**.

The quality of the receiver radio-frequency chain (RFC) and the regular variations in the signal level at the reception point play an important role in the potential accuracy of the signal phase measurements. In particular, the sustainable phase tracking threshold equals 15° [17] is almost reached under conditions of the worst radio-frequency chain parameters (**Table 2**) and the minimum signal receiving

Noise source	Noise temperature, K	Power spectral density dBW/Hz
Preamplifier noise	120 ... 300	−204 ... −208
Antenna noise	30 ... 60	−214 ... −211
Sky noise (all sources)	100	−208
Earth noise	1.2 ... 3	−228 ... −224
Sun noise	0.06	−241
Total value	251.26 ... 463.06	−204.6 ... −202.1

Table 1.
The receiver and external thermal noises.

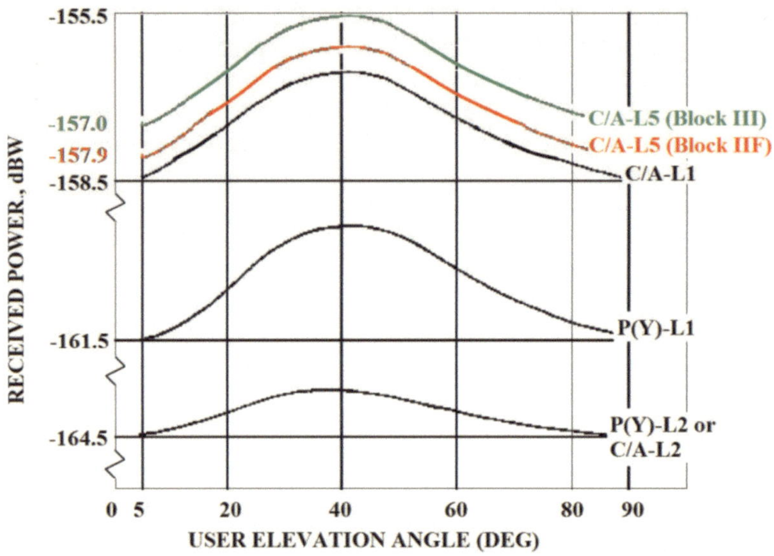

Figure 1.
The power levels of signals received by a linearly polarized antenna with 3dBi antenna gain at L1, L2, and L5 GPS frequencies (the curves were reconstructed based on [19, 20]).

Frequency, MHz	Minimal value σ_φ, deg	Maximal value σ_φ, deg
L1 = 1575.42	1.59	7.22
L2 = 1227.60	3.35	14.85
L5 = 1176.45	1.33	6.06

Table 2.
Noise values of phase measurements.

level at the L2 frequency. Thus, although the phase measurements yield the best accuracy for ionospheric scintillation detection, still the careful presetting of GNSS receiver hardware and the consideration of measurement conditions are needed. To note, under the similar conditions, the best accuracy of the phase measurements is achieved if the signals are used at the L5 GPS frequency. This can be explained by the highest carrier-to-noise ratio in the given measurement channel (**Figure 1**).

Another important factor for the high accuracy of carrier phase measurements is the correct choice of the PLL settings such as accumulation time (T_{COR}) and the PLL filter noise bandwidth (ΔF_{PLL}). It is known that the third-order tracking system has stable and unstable operation zones. If there are no impacts on the navigation receiver in the form of vibrations, jerks, and electromagnetic jammer interference, then the stable tracking of the carrier phase is provided with the following conditions fulfilled [17]:

$$0 < \Delta F_{PLL} < \frac{0.7}{T_{COR}} \tag{11}$$

When using an optimal phase discriminator, the measured parameter (phase) should not be changed during the accumulation time (T_{COR}). In this case, when estimating the phase, it is necessary to head for the longest character of the transmitted message. This is the character of the navigation message, which is transmitted simultaneously with the ranging code on the same carrier frequency. If the duration of the navigation message character is 20 ms [21], the accumulation of measurements should be $T_{COR} \geq 20$ ms.

The measured parameter is not obligatory constant within T_{COR} interval in case of a quasi-optimal phase discriminator [21]. Therefore, the accumulation time should be within the interval $1ms \leq T_{COR} \leq 20ms$. Here, the ranging code sequence length (1 ms for the CA code) determines the lower limit of the accumulation time (1 ms). The final decision about the optimal T_{COR} value is limited by two factors: (1) the carrier-to-noise ratio in the phase measuring channel and (2) the influence of the low-frequency processes on the phase measurement accuracy. The longer the time interval, the higher both the carrier-to-noise ratio and the phase measurement accuracy. However, with an increase of the accumulation time more than 10–20 ms, the effects of instability in the reference oscillator frequency and Doppler frequency drift can appear [21]. Therefore, it is not appropriate to increase the accumulation time over these limits.

After the determination of the optimal T_{COR} value, the selected PLL noise bandwidth should satisfy Eq. (11). In addition, according to Eq. (1) the noise level of the phase measurement depends on the noise bandwidth ΔF_{PLL}. Therefore, the practical choice of the noise bandwidth depends on the expected measurement conditions and usually lies within the range from 10 to 20 Hz. If there is an impact of external electromagnetic interference, the phase tracking stability reduces. Therefore, the choice of the wider noise bandwidth increases the reliability of the phase tracking. Finally, according to expressions (3) and (4), the increase of the noise bandwidth leads to the proportional increase of the average RMS of the receiver equipment thermal noise. On the other hand, as ΔF_{PLL} increases, the noise component related to the short-term frequency instability of the reference oscillator decreases. Thus, the noise bandwidth can be reduced without the significant loss of the phase measurement quality by using a better-quality reference oscillator.

The multipath effect is another important source of the carrier phase noises. In general, the phase error due to multipath can be calculated as a difference between the carrier phase of the reflected composite signal and the carrier phase of the direct signal. In the presence of multipath propagation, the composite signal phase shifts randomly from the direct signal phase, and the NCO-generated local carrier locks to the composite carrier phase, resulting in the error of the phase measurement. In the case of one reflected signal, the error of the phase measurement is defined as follows [10]:

$$\Delta\Psi = arctan\left(\frac{\alpha_1 \bullet R(\hat{\tau_C} - \tau_1) \bullet sin\,\varphi_1}{R(\hat{\tau_C}) + \alpha_1 \bullet R(\hat{\tau_C} - \tau_1) \bullet cos\,\varphi_1}\right) \tag{12}$$

where $R(\hat{\tau_C})$ is the autocorrelation function of the PRN code, $R(\hat{\tau_C} - \tau_1)$ is the cross-correlation function between the direct GNSS signal and the reflected signal, $\hat{\tau_C}$ is the receiver estimate of the incoming signal code delay, τ_1 is the reflected signal code delay, α_1 is the reflection coefficient that corresponds to the Signal to Multipath Ratio (SMR) as $SMR = 20 \bullet log\left(\alpha_1^{-1}\right)$, and φ_1 is the phase of the reflected signal.

If the direct signal has no distortion in the form the PRN code, the autocorrelation function ($R(\hat{\tau_C})$) depends on the front-end bandwidth of the GNSS receiver radio-frequency chain. The PRN codes have one main lobe and several side lobes in the frequency domain. In practice, the signal is band limited, and only the main lobe and one or more side lobes are used for the signal processing. As a result, the sharp correlation peaks are rounded and the ends are trailed-off. It was found earlier that the RFC bandwidth affects the maximum error value significantly [10, 11]. In the case of the unlimited bandwidth, the misalignment in the $\hat{\tau_C}$ value computation is zero. In the case of 10 MHz bandwidth, the misalignment is not equal to zero and can vary within $\pm 0.03\, t_C$, where t_C is PRN code chip length. The narrower bandwidth of 2 MHz brings the significant misalignment to the calculation of the $\hat{\tau_C}$ value which can reach the values of $\pm$ (0.1 ... 0.3) t_C [10].

The cross-correlation function $R(\hat{\tau_C} - \tau_1)$ significantly depends on the early-late correlator spacing (d) and PRN code rate. It is well known that the code delay discriminator output ($\Delta_{d,out}$) depends on the correlator spacing time (Td), the input tracking error ($\hat{\tau_C} - \tau_c$), and the PRN code chip length (t_C) as follows [17]:

$$\Delta_{\mathrm{d,out}} = -2 \bullet \frac{(\hat{\tau_C} - \tau_c)}{t_C}_{\tau_e \leq Td} \tag{13}$$

This equation describes the discriminator output in case if the input tracking error (τ_e) is within linear part of the discriminator performance. The maximal discriminator output value is limited by the correlator spacing time and depends on the code chip length [17]

$$\Delta_{\mathrm{d,MAX}} = 2 \bullet \frac{T_d}{t_C}_{\tau_e = Td} \tag{14}$$

Thus, both the correlator spacing and the PRN code chip length define the maximal code tracking error value and, as a result, the cross-correlation function $R(\hat{\tau_C} - \tau_1)$. Let us consider the particular example of L1 C/A code and the coherent discriminator using a standard correlator with the correlator spacing of $Td = \pm 0.5 t_C$ and a narrow correlator with the spacing of $Td = \pm 0.1 t_C$. The work [10] proved that for this particular case, the maximum and the minimum errors are much higher than the narrow correlator with 0.1 chip spacing. If the correlator spacing is $\pm 0.5 t_C$, the code tracking error lies within $\pm 0.4 \ldots 0.5\, t_C$. It corresponds to the code delay computation error $\hat{\tau_C} = \pm$ 120 ... 150 m for C/A PRN code. In contrast to that, in the case of the narrow correlator, the code tracking error is $\hat{\tau_C} = \pm$ 25 ... 30 m for C/A code.

To estimate the possible impact of the PRN code rate on the multipath error, the multipath error envelopes can be used [22, 23]. **Table 3** illustrates the GPS PRN code characteristics transmitted at L1 and L5 GPS frequencies. **Table 4** was reconstructed based on the results [23]. It illustrates the maximal code multipath error ($\hat{\tau_C} \bullet c$) in relation to the PRN code rate and correlator spacing for the coherent discriminator.

Table 4 shows that the size of the area, where the multipath effect is significant, depends on the code rate or, to be exact, on the PRN chip length (t_c). This is because the discriminator function value varies within $0 < Td < t_c$. If the input code tracking error exceeds t_c, the discriminator function values saturate. Hence, the code discriminator is sensitive within $\pm\, t_c$. Moreover, in the case of the standard correlator, the multipath error beyond the multipath delay of 1.5 t_c can differ from zero. This is due to the fact that the PRN code autocorrelation characteristics has one major peak and many minor peaks [17]. If the reflected signal is received more

Frequency/PRN code	Carrier frequency, MHz	Code rate (Mbps)
L1 C/A	1575.25	1.023
L5 I5, L5 Q5	1176.45	10.23

Table 3.
L1, L2, and L5 GPS signal characteristics [19, 20].

Frequency/PRN code	Maximal code multipath error ($\hat{\tau_C} \bullet c$), meters		Maximal relative multipath delay ($\tau_1 \bullet c$), meters	
	Td = $\pm 0.5t_C$	Td = $\pm 0.1t_C$	Td = $\pm 0.5t_C$	Td = $\pm 0.1t_C$
L1/CA	39.0	8.0	350	300
L5/I5, L5/Q5	4.0	0.8	40	35

Table 4.
The code multipath error in relation to the relative multipath delay at the fixed SMR = 3 dB and different correlator spacing.

than 1.5 chip delayed, it can cause the minor peak or a non-zero correlation value as well [24]. This effect is not significant for our analysis, thus we will not consider it further.

The maximum error values of the phase measurement due to multipath are calculated according to Eq. (12). It was supposed that there is only one reflected signal that has the phase shift angle $\varphi_{1,max}$, rad and delayed τ_1 seconds. This angle corresponds to the case when the multipath errors reach the maxima and affects the multipath error envelope which contains all the smaller variations of the $\Delta\Psi$ values. The angle $\varphi_{1,max}$ can be found by differentiating Eq. (12) with respect to φ_1, putting it to zero and solving it for φ_1. It results in the following [10]:

$$\varphi_{1,max} = cos^{-1}\left(\frac{-\alpha_1 \bullet R(\hat{\tau_C} - \tau_1)}{R(\hat{\tau_C})}\right) \tag{15}$$

Figure 2 shows the standard deviations of the carrier phase multipath errors with respect to multipath delays for different SMR using the correlator spacing of $Td = \pm 0.1t_C$ and the coherent discriminator for code tracking. The results in **Figure 2** correspond to L1 C/A PRN code (solid lines) and L5 I5 (Q5) PRN codes (dashed lines). Here, the unlimited RFC bandwidth and τ_1 variations within the range of $\pm t_c$ are supposed. The results are obtained changing the reflected signal relative phase shift by discrete steps of 0.1 of a total phase cycle, calculating the

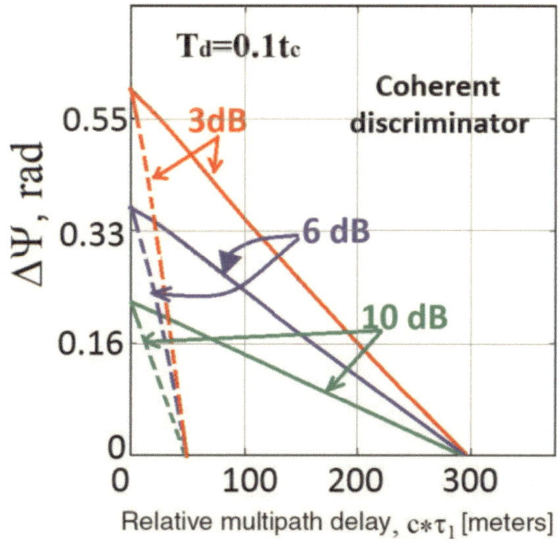

Figure 2.
The multipath error envelopes for L1 C/a (solid lines) and L5 I5 (Q5) PRN codes (dashed lines) in relation to SMR.

multipath error ($\Delta\Psi$) at each step and then taking their mean values and standard deviation.

According to **Figure 2,** there is a dependence of the error on SMR. The magnitude of the multipath error ($\Delta\Psi$) is proportional to the strength of the multipath signal. Moreover, the multipath error value is independent on the carrier wavelength (Eq. 12), but it is mostly a function of the antenna-reflector distance through the correlation function $R(\hat{\tau_C} - \tau_1)$. If the multipath delay (τ_1) is high, the correlation value decreases and so does the multipath error amplitude. The maximal multipath error in the phase measurement does not exceed 0.6 rad under the above mentioned assumptions (**Figure 2**). However, under real conditions the multipath error is formed as a sum of several reflected signals or as a result of another kind of multipath sources such as diffuse scattering or diffraction. Thus, the higher values of the error of the phase measurement due to multipath can be expected. The authors [10, 11] demonstrated that the maximal value of the error due to multipath does not depend significantly on the code correlator spacing and there is no similar dependence on the code discriminator type as well.

3. Experimental results and analysis

3.1 Indices comparison

This section discusses the performance of the "standard" ionospheric scintillation indices and the index D2fi based on high-rate sampling data. The D2fi index and the ionospheric indices/parameters *TEC, DROTI, S4*, and $\sigma\varphi$ were compared during the geomagnetic storm conditions. The ionospheric scintillations are considered to be more typical for high and low latitudes. Mid-latitude scintillations are supposed to occur much less often. Here, first, we analyze the data of the mid-latitude station where the scintillation detection is a rather challenging problem and estimate the indices performance. Then we consider the example of high latitudes.

The 50 Hz *L1* and *L2* GPS data were obtained at the mid-latitude station ISTP (Irkutsk, Russia, geographic coordinates 52° N, 104° E) equipped with JAVAD GNSS receiver. The station is a part of SibNet network [25, 26].

As the de-trended TEC data is used to calculate *DROTI* indices, the uncalibrated code-leveled phase TEC time series were derived from GPS phase and code measurements for this study. The phase TEC time series were de-trended by the centered moving window with 30 second accumulation time. *DROTI* values were calculated from the de-trended 50 Hz TEC data with 1 second time resolution based on [5]. The indices *S4* and $\sigma\varphi$ were calculated from the de-trended 50 Hz *L1* data based on the standard procedure [6] with 1 second time resolution as well.

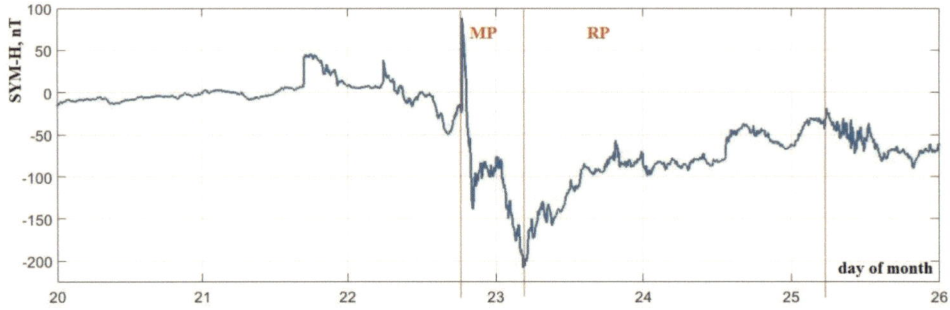

Figure 3.
SYM-H variations during June 20–25, 2015. The MP and RP are indicated with the vertical red lines.

The storm period was chosen for the analysis as geomagnetic storms are known to cause ionospheric disturbances including the small-scale disturbances that are of the particular interest for this work. The intense storm of June 22–25, 2015, was under analysis. **Figure 3** shows SYM-H index variations during the storm. Main phase (MP) and recovery phase (RP) of the storm are marked with red lines [27]. SYM-H reached its minimum on June 23rd. SYM-H index data was obtained from Data Analysis Center for Geomagnetism and Space Magnetism following the link http://wdc.kugi.kyoto-u.ac.jp/aeasy/index.html (last access: August 2018).

According to [3], the relationship between *S4*, $\sigma\varphi$, and *DROTI* is complex, but in most cases the *S4* increase means *DROTI* increase and vice versa. **Figure 4** shows variations of the D2fi index, *DROTI*, *S4*, and $\sigma\varphi$ indices during the storm for the GPS satellites PRN 04, PRN 15, and PRN 27 observed at ISTP station. Good

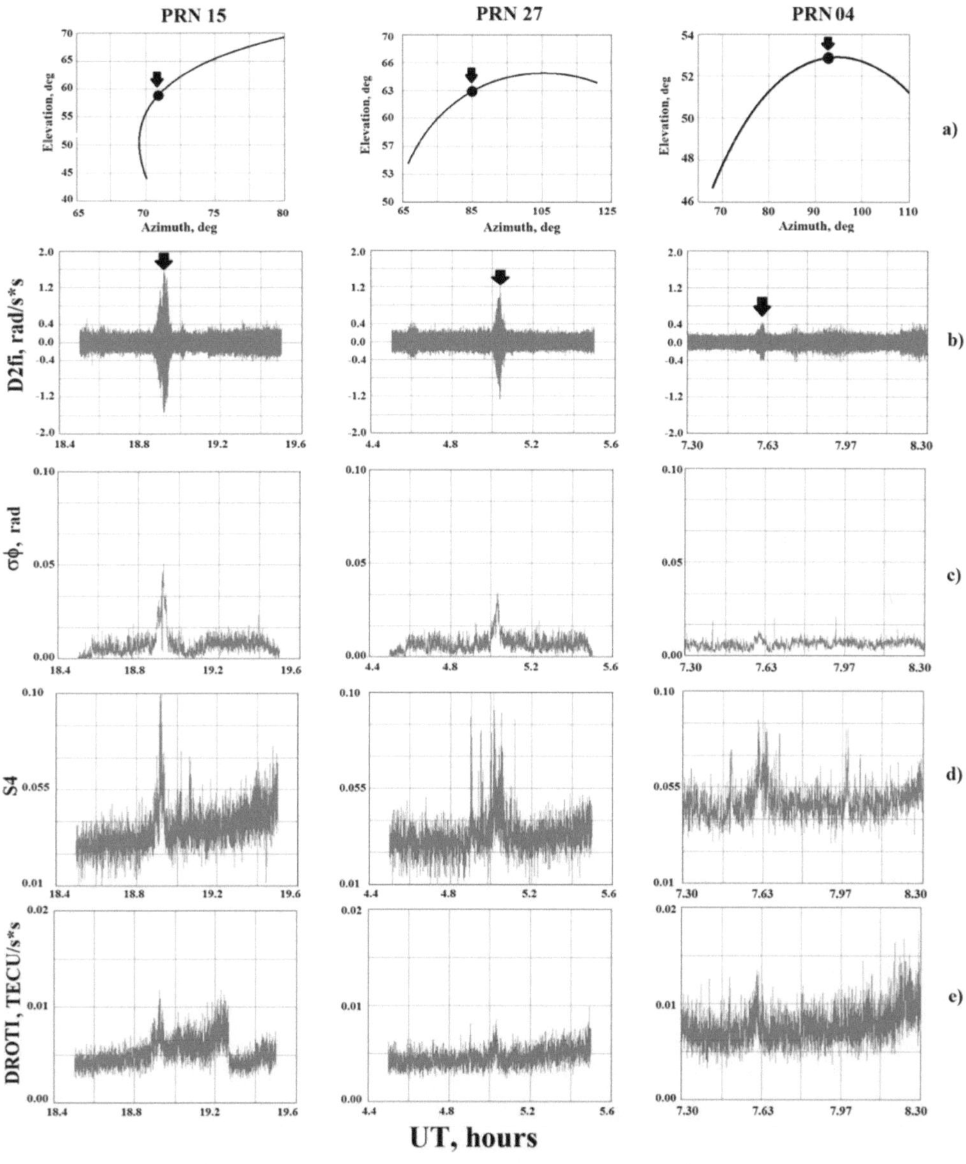

Figure 4.
Time behavior of D2fi and standard scintillation indices. The dots indicate the approximate SV angular positions when the scintillation events were observed. (a) Elevation of the satellites, (b) D2fi, (c) $\sigma\varphi$, (d) S4 and (e) DROTI.

correlation between the D2fi index and $\sigma\varphi$ variations is seen for all the scintillation events and for all considered satellites, including the weakest event for PRN 04 (**Figure 4b** and **c**). Nevertheless, the peaks of the D2fi index are pronounced more sharply for all the cases. The correlation between the D2fi index and *S4* variations is worse. There is a general similarity in behavior of these parameters, but *S4* distribution is rather noisy and contains several peaks which do not coincide in time with the peaks of the D2fi index (**Figure 4b** and **d**).

The worst correlation is between the D2fi index and *DROTI* for all the cases under consideration (**Figure 4b** and **e**). The form of *DROTI* envelope significantly differs from the envelope of the D2fi index. To add, *DROTI* observations are rather noisy. Almost no *DROTI* response is seen for the SV PRN 27 (**Figure 4e**, middle panel). The small-scale ionospheric irregularities do not provoke significant TEC response [28]. Consequently, even weaker response can be expected in TEC-derived indices like *DROTI*, which is probably the case of **Figure 4e**.

Let us consider the advantages of the D2fi index in comparison to other indices by the example in **Figure 4**. First, it marks the sharper and more precise in time response to small-scale turbulences than other indices. Furthermore, only one GPS frequency is needed to obtain the D2fi index. Thus, it avoids the possible impact from the inter-frequency noises and *L1*-aiding technique features [23]. Third, as the D2fi index is calculated from either *L1* or *L2* phase data, it does not require any additional preprocessing and does not depend on the data processing technique [8]. Finally, another advantage of the D2fi index is its high sensitivity. We recall that mid-latitudes are usually considered as the region where the scintillation occurrence is null except during geomagnetic disturbances. Even for the presented case, the scintillation intensity is very low (S4 is not higher than 0.1, **Figure 4d**). Nevertheless, the D2fi index response on these scintillation events is clear, and it is more precise in time than other scintillation indices under consideration.

Now, let us consider the data from high-latitude region, where scintillations are more frequent. **Figures 5** and **6** are similar to **Figure 4** and show the results derived from the 50-Hz data at stations EDM (53,35° N, 247,02° W) and GJO (68,63° N, 254,15° W). Both stations belong to the Canadian High Arctic Ionospheric Network (CHAIN) [29] and equipped with the same type SEPTENTRIO PolaRxS GNSS receivers [30]. The station EDM is still within mid-latitudes (however in Canada it strictly depends on current geomagnetic conditions), but the station GJO is in high-latitude region.

The scintillation events are detected at both sites in the same time interval by all the considered indices: at EDM with PRN 30, PRN 26, and PRN 15 (**Figure 5**) and at GJO with PRN 06 (**Figure 6**).

It is seen that the weaker scintillation, the weaker the response of D2fi and $\sigma\varphi$, which is not surprising as both indices are calculated from the same phase ranging data. Note that $\sigma\varphi$ index quality depends on the phase de-trending and filtering procedure. This could bring the artificial effect that is seen at 19.87 UT at **Figure 6c** (left column).

The comparison of different indices allows us to reveal the prevalence of phase or amplitude scintillations. In our case (**Figure 5**) the obvious difference in S4 and $\sigma\varphi$ behavior is seen for PRN 30, PRN 26, and PRN 15. Amplitude scintillations prevail at the ray path from PRN 30 (S4 exceeds 0.15, **Figure 5c**). On the other hand, phase scintillations are predominant at the ray paths from PRN 26 and 15 ($\sigma\varphi$ index reaches 0.2 but S4 index does not exceed 0.02 at the same time).

To sum up, **Figures 4–6** prove the following: (a) D2fi peaks are caused by scintillation events (as there are also responses in other scintillation indices though less precise) and (b) that the D2fi index shows more sensitivity to phase scintillations.

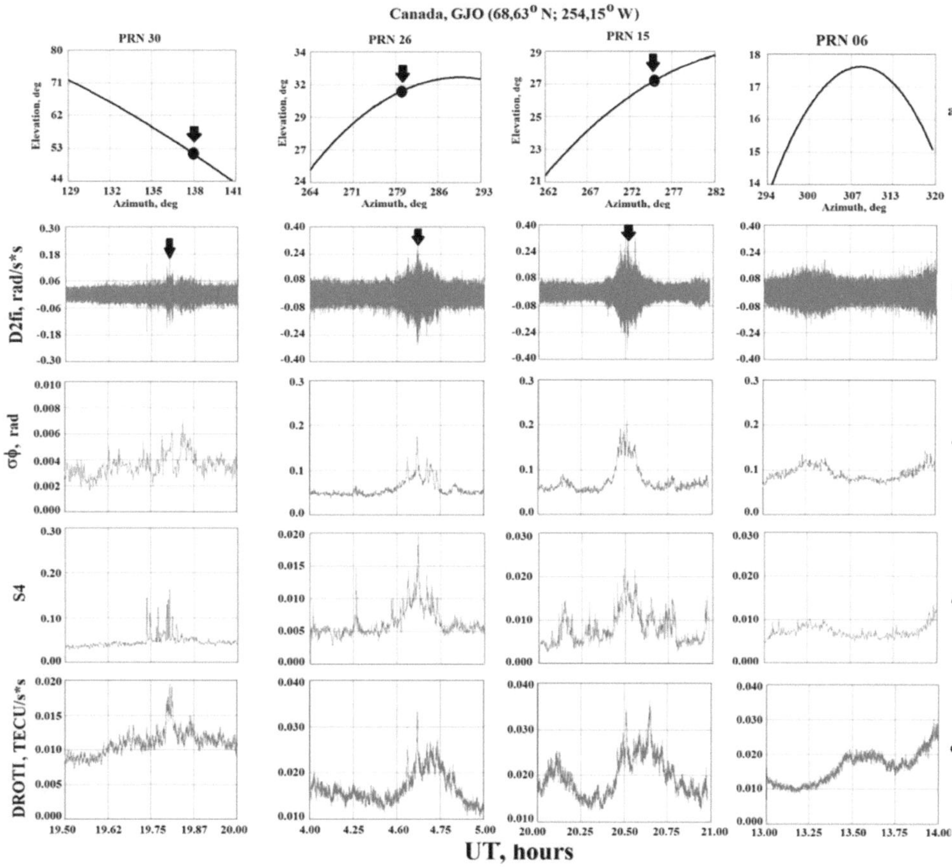

Figure 5.
The same as in **Figure 4**, *but for the high-latitude GJO station (Canada).*

3.2 Time resolution comparison

The time resolution of input data is very important to detect scintillations. For example, the work [12] showed the significant sampling rate influence on *ROTI*. Indeed, the minimal size of the refractive irregularities is about the first Fresnel zone size (300–400 m at GNSS frequencies band). Such irregularities can cause both refractive and diffractive variations not only in the input carrier phase data but also in the ionospheric TEC and its derivatives as well as variations of *S4* and $\sigma\varphi$ indices which are calculated based on 0.1–10 Hz data.

The smaller irregularities (from tens of meters to 100–300 m) are mostly considered to provoke the diffractive amplitude and phase variations. To detect them the highest time rate possible is needed (higher that 10 Hz). Diffractive phenomena can cause the phase scintillations that are usually accompanied by the intense amplitude fluctuations. These can be detected by $\sigma\varphi$ and *S4* indices. When the diffractive Fresnel irregularities dominate, CNR and/or *S4* can vary significantly and show high correlation with $\sigma\varphi$ [3].

Several kilometer size irregularities usually cause the refractive scintillations of 0.01–0.1 Hz. When such irregularities dominate, *S4* does not vary significantly and almost has no correlation to *ROTI*, *DROTI*, and even to $\sigma\varphi$. Scintillations of refractive origin are better observed with the sharp TEC variations (i.e., by means of ROTI and DROTI) and with $\sigma\varphi$ [3, 5]. There are studies focused on the scintillation

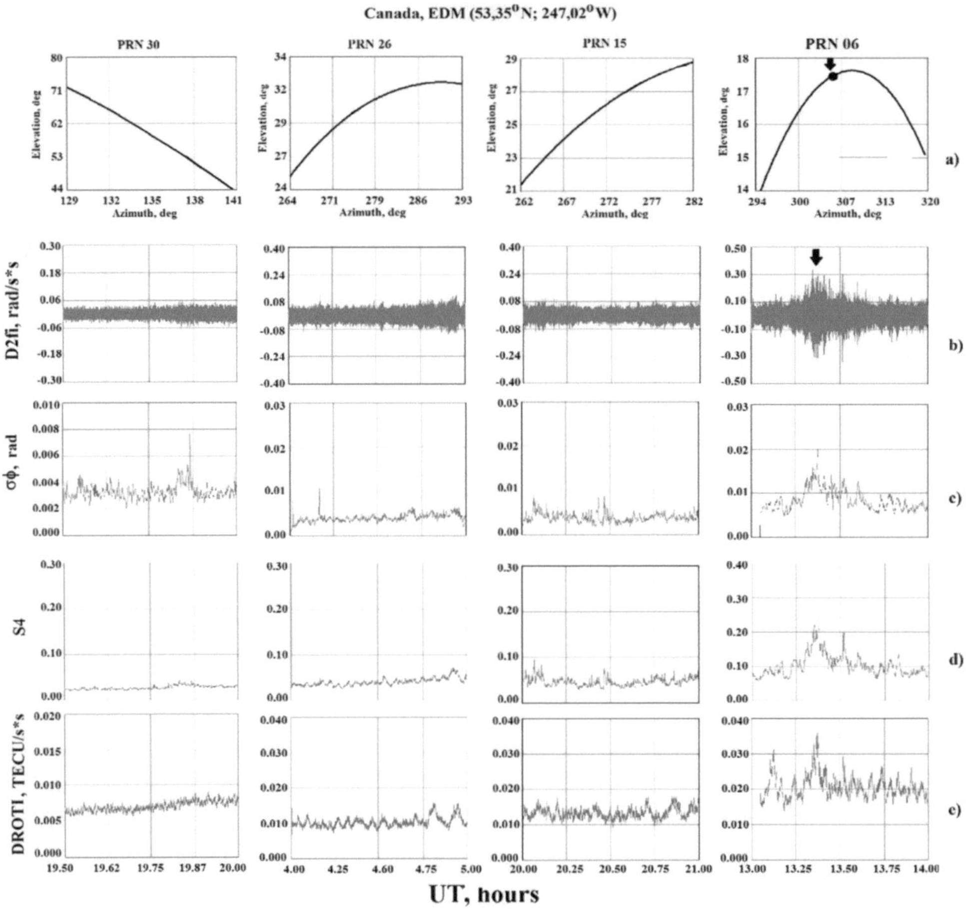

Figure 6.
The same as in ***Figure 4****, but for the high-latitude EDM station (Canada).*

indices use based on the data of high-latitude receivers. For instance, the sDPR index was introduced in [15].

Usually, the irregularities of different scales are present in the ionosphere simultaneously**.** It can occur during the volcanic eruptions, powerful explosions, rocket launching**,** under disturbed geomagnetic conditions, etc. [4]. The ionospheric irregularities can move with the quiet different velocities and in different directions. The 1 Hz or lower time resolution data does not allow us to reveal if the ionospheric event was caused by the diffractive irregularities of hundreds of meters or by the larger refractive irregularities of tens of kilometers.

We suggest that the high data sampling rate such as 10 Hz and higher provides the opportunity to reveal and analyze the weak small-scale ionospheric irregularities. To test this assumption, we compared 1, 10, and 50 Hz time series of the D2fi index for the same events and under the same geomagnetic storm conditions. **Figure 7** shows the results of comparison for PRN 04, PRN15, and PRN27 at ISTP station on June 22, 2015, during the main phase of the geomagnetic storm (**Figure 3**).

The D2fi index obtained from 1 Hz GPS data does not reveal any scintillation event for all three satellites (**Figure 7c**). In contrast**,** the time series obtained from 10 Hz data show the clear peaks for the SV PRN 15 and PRN 27 (**Figure 7b**), but not for the weakest event for SV PRN 04 (**Figure 7b**, left). The peaks of 50 Hz time series are the most pronounced for all the satellites (**Figure 7a**). Note that the 1 Hz

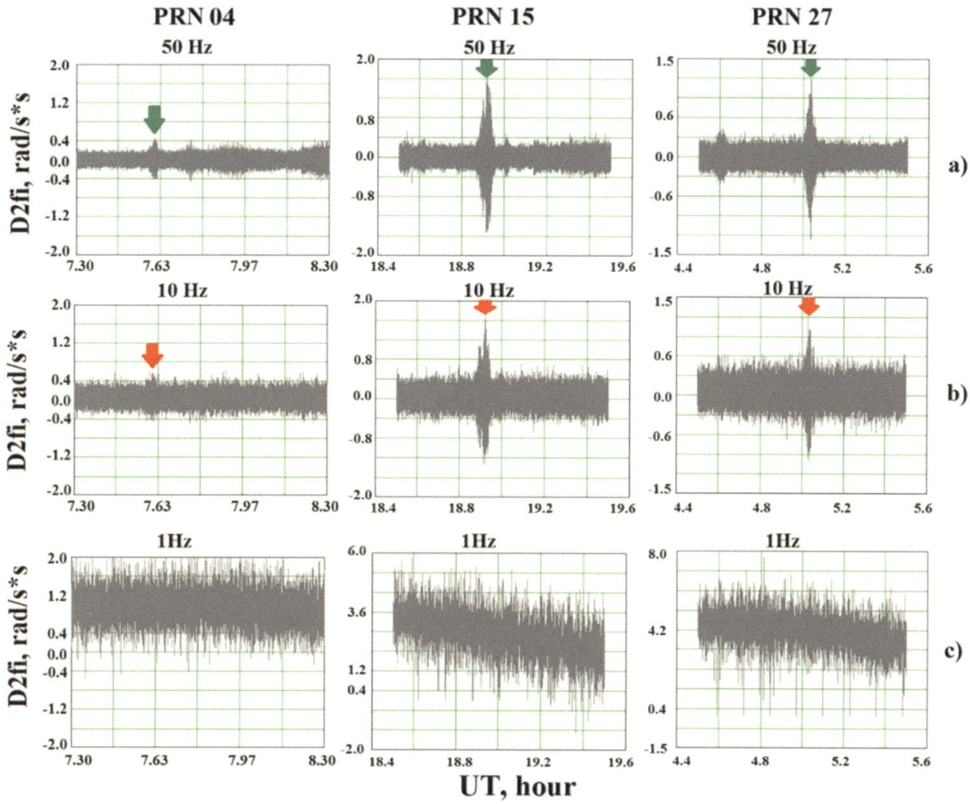

Figure 7.
The D2fi index in case of 50 Hz data sampling rate (a), 10 Hz data sampling rate (b), and 1 Hz data sampling rate (c) for PRN 04, PRN 15, and PRN 27 on June 22, 2015, at ISTP station.

data shows both the highest noise level and the additional regular trend. The low-frequency trends are mostly removed from the time series of higher sampling rate.

In case of the highest data rate (50 Hz), the background values of D2fi do not exceed 0.4 rad/s*s (**Figure 7a**). For the lower data rate (10 Hz), the weak regular trend appears, and the background noise increases to 0.6 rad/s*s (**Figure 7b**). The D2fi variations increase 4–5 times and exceed 2–3 rad/s*s in the last case (1 Hz data, **Figure 7c**).

Apart from the ionospheric scintillations, one of the common sources of the phase fluctuations is the multipath effect. The majority of the multipath-induced fluctuations are observed at lower elevation angles. It is also not a thorough determination of multipath as it is possible to observe it at the higher elevations as well [1]. Thus, we should test if the scintillation events revealed above are related to multipath and/or blocked signal effects. Usually, the multipath-induced phase variations are caused by the repeating events due to local reflection or diffuse scattering. The picture of such events repeats from day to day at the same location. At the same time, the picture of such "scintillations" has the regular time shift about 16 s from one day to another due to GPS orbits daily motion [17]. This means that to determine whether the scintillation candidate events are caused by repeating local multipath effects, the raw data for the day before and after the scintillation should be analyzed. **Figure 8** illustrates such the analysis for 50 Hz data on June 21, June 22, and June 23, 2015, for PRN 04, PRN15, and PRN27.

No significant phase scintillations on the day before (June 21, **Figure 8**, left column) and/or after (June 23, **Figure 8**, right column) were observed. In contrast, there were the sharp and rapid variations of the second-order derivative of the

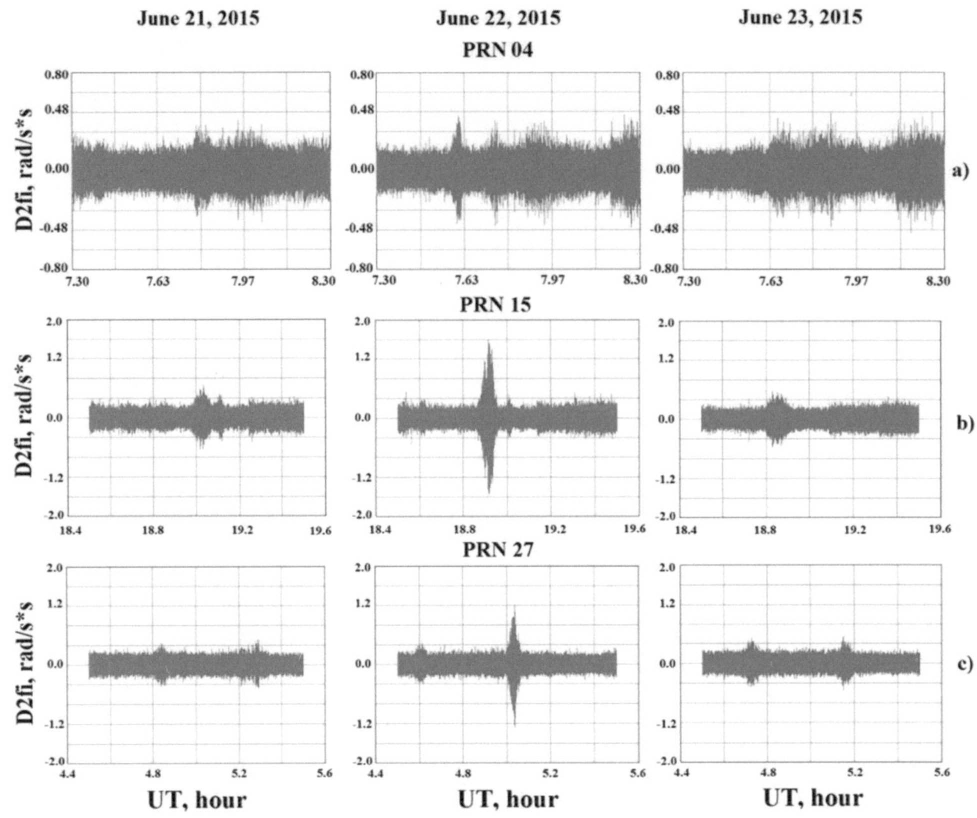

Figure 8.
The D2fi index during June 21–23, 2015, for satellites PRN04 (a), PRN15 (b), PRN27 (c).

carrier phase on June 22, 2015, for all the satellites. This fact proves that the phase scintillation events observed on June 22, 2015, are not related to the multipath effect. Thus, the above mentioned phase scintillation events probably have the ionospheric origin.

4. Conclusions

The performance of the well-known "standard" ionospheric scintillation indices ROTI, DROTI, $S4$, $\sigma\varphi$, and the new scintillation index D2fi that is the second-order derivative of the GPS signal carrier phase was analyzed in this study. The features of GNSS receivers and antennas that can have an effect on this performance were considered. The benefits and limitations of the indices were discussed.

The overall accuracy of the GNSS carrier phase measurements is limited by both thermal and external noises and significantly depends on the GNSS hardware and software presets and architecture. The accuracy of the carrier phase measurements can be improved if the particular specification is used for GNSS equipment suggested for the ionospheric studies. This particular specification means that the narrowband code delay discriminator, the large code rate for the open-access GNSS signals, the expanded front-end bandpass of the RFC, the low-noise preamplifier, and the specific pattern antenna should be specified for the ionospheric study.

In the present study, the new index D2fi is proved to be an effective tool to detect the small-scale ionospheric irregularities. It was shown that the sensitivity of the D2fi index depends on the data sampling rate. The higher the sampling rate, the clearer the peaks of the D2fi index, and the weaker both the noise background and the low-frequency trend. The comparison between the D2fi index and DROTI, *S4*, and $\sigma\varphi$ showed that they have different "sensitivities." Each index has its own "critical" sensitivity for the particular ionospheric turbulences depending on the data sampling rate and preprocessing procedures. The advantage of the new D2fi index is that it is easily derived from the single-frequency carrier phase data. It provides both the reliable detection of the ionospheric scintillation and the phase time series de-trending with no complex data preprocessing. The new index can be applied as an independent scintillation indicator or as an additional tool together with other scintillation indices.

Acknowledgements

This work was supported by the Russian Federation President Grant No. MK-3265.2019.5 and by grant No. 18-05-00343 from the Russian Foundation for Basic Research. LANCE acknowledges partial support from CONACyT LN-299022, CONACyT PN 2015-173, and CONACyT-AEM Grant 2017-2101-292684. ISTP staton data were recorded by the Angara Multiaccess Center facilities (http://ckp-angara.iszf.irk.ru) at ISTP SB RAS within the base financing of FR program II.16.

Conflict of interest

Vladislav Demyanov is the principal author and the corresponding author of this book chapter. The text and the figures presented in this book chapter were not published anywhere else before.

Author details

Vladislav V. Demyanov[1*], Maria A. Sergeeva[2,3]
and Anna S. Yasyukevich[1]

1 Laboratory of Elaboration of New Methods for Atmosphere Radio Diagnostics, Institute of Solar-Terrestrial Physics, Irkutsk, Russia

2 SCiESMEX, LANCE, Instituto de Geofisica, Unidad Michoacan, Universidad Nacional Autonoma de Mexico, Michoacan, Mexico

3 CONACYT, Instituto de Geofisica, Unidad Michoacan, Universidad Nacional Autonoma de Mexico, Michoacan, Mexico

*Address all correspondence to: vv.emyanov@gmail.com

IntechOpen

© 2019 The Author(s). Licensee IntechOpen. This chapter is distributed under the terms of the Creative Commons Attribution License (http://creativecommons.org/licenses/by/3.0), which permits unrestricted use, distribution, and reproduction in any medium, provided the original work is properly cited.

References

[1] McCaffrey AM, Jayachandran PT. Spectral characteristics of auroral region scintillation using 100 Hz sampling. GPS Solutions. 2017;**21**:1883-1894. DOI: 10.1007/s10291-017-0664-z

[2] Fanis M, Stathis S. PLL bandwidth and noise in 100 Hz GPS measurements. GPS Solutions. 2015;**19**:173-185. DOI: 10.1007/s10291-014-0378-4

[3] Bhattacharrya A, Yen KC, Franke SJ. Deducing turbulence parameters from transionospheric scintillation measurements. Space Science Reviews. 1992;**61**:335-386

[4] Afraimovich EL, Perevalova NP, editors. GPS-Monitoring of Earth Upper Atmosphere. Irkutsk: Russian Academy of Sciences, Siberian Branch; 2006. 460p. ISBN 5-98277-033-7

[5] Pi X, Mannucci AJ, Lindqwister UJ, Ho CM. Monitoring of global ionospheric irregularities using the worldwide GPS-network. Geophysical Research Letters. 1997;**24**:2283-2286. DOI: 10.1029/97GL02273

[6] Van Dierendonck AJ, Klobuchar J, Hua Q. Ionospheric scintillation monitoring using commercial single frequency C/A code receivers. In: Proceedings of ION GPS 1993; Salt Lake City, Utah. 1993. pp. 1333-1342

[7] Ghoddousi-Fard R. Impact of receiver and constellation on high rate GNSS phase rate measurements to monitor ionospheric irregularities. Advances in Space Research. 2017; **60**(9):1968-1977

[8] Mushini SC, Jayachandran PT, Langley RB, MacDougall JW, Pokhotelov D. Improved amplitude-and phase-scintillation indices derived from wavelet detrended high-latitude GPS data. GPS Solutions. 2012;**16**(3):363-373

[9] Priyadarshi S, Zhang QH, Thomas EG, Spogli L, Cesaroni C. Polar traveling ionospheric disturbances inferred with the B-spline method and associated scintillations in the southern hemisphere. Advances in Space Research. 2018;**62**(11):3249-3266

[10] Jayanta KR. Mitigation of GPS code and carrier phase multipath effects using a multi-antenna system [thesis]. Calgary, Alberta: The University of Calgary; 2000

[11] Qiongqiong J, Renbiao W, Wenyi W, Dan L, Lu W, Jie L. Multipath interference mitigation in GNSS via WRELAX. GPS Solutions. 2017;**21**:487-498. DOI: 10.1007/s10291-016-0538-9

[12] Jacobsen KS. The impact of different sampling rates and calculation time intervals on ROTI values. Journal of Space Weather Space Climate. 2014;**4**: A33. DOI: 10.1051/swsc/2014031

[13] Ledvina BM, Makela JJ, Kintner PM. First observations of intense GPS L1 amplitude scintillations at multitude. Geophysical Research Letters. 2002;**29** (14):4. DOI: 10.1029/2002GL014770. Available from: https://agupubs.onlinelibrary.wiley.com/doi/epdf/10.1029/2002GL014770

[14] Kintner PM, Kil H, de Paula E. Fading time scales associated with GPS signals and potential consequences. Radio Science. 2001;**36**(4):731-743

[15] Ghoddousi-Fard R, Prikryl P, Lahaye F. GPS phase difference variation statistics: A comparison between phase scintillation index and proxy indices. Advances in Space Research. 2013;**52**(8):1397-1405

[16] Pavelyev AG, Liou YA, Wickert J, Schmidt T, Pavelyev AA. Phase

acceleration: A new important parameter in GPS occultation technology. GPS Solutions. 2010;**14**:3-11

[17] Kaplan ED. Understanding GPS: Principles and Applications. Boston, USA and London, UK: Artech House; 1996. p. 556

[18] MOPS Based Procedure for Minimum Recommended Testing of LightSquared RFI to GPS Aviation Receivers: Appendix A. 2017; p. 220. Available from: https://ecfsapi.fcc.gov/file/7021690472.pdf

[19] Global Positioning Systems Directorate Systems Engineering and Integration Interface Specification: IS-GPS-200J. 2018. Electronic resource. Available from: https://www.gps.gov/technical/icwg/IS-GPS-200J.pdf

[20] Global Positioning System Wing Systems Engineering and Integration Interface Specification. IS-GPS-705, Revision A: Navstar GPS Space Segment/User Segment L5 Interfaces. 2010. Electronic resource. Available from: https://www.gps.gov/technical/icwg/IS-GPS-705A.pdf

[21] Tsui JB. Fundamentals of Global Positioning System Receivers : A Software Approach. 2nd ed. Hoboken, New Jersey, USA: John Wiley & Sons, Inc., Publication; 2005. 605p. ISBN 0-471-70647-7

[22] Van Nee R. Multipath effects on GPS code phase measurements. In: Proceedings of ION GPS. 1991. pp. 915-924

[23] Padma B, Kai B. Performance analysis of dual-frequency receiver using combinations of GPS L1, L5, and L2 civil signals. Journal of Geodesy. 2019;**93**:437-447. DOI: 10.1007/s00190-018-1172-9

[24] Braasch MS. GPS multipath model validation. In: Proceedings of ION PLANS. 1996. pp. 672-678

[25] Yasyukevich YV. 2017. The 50 Hz JAVAD Data Set for the Case Study. Available from: 10.5281/zenodo.848325 [Accessed: 30 May 2019]

[26] Yasyukevich YuV, Vesnin AM, Perevalova NP. SibNet–Siberian global navigation satellite system network: Current state. Solar-Terrestrial Physics. 2018;**4**(4):63-72. DOI: 10.12737/stp-44201809

[27] Piersanti M, Alberti T, Bemporad A, Berrilli F, Bruno R, Capparelli V, et al. Comprehensive analysis of the geoeffective solar event of 21 June 2015: Effects on the magnetosphere, plasmasphere, and ionosphere systems. Solar Physics. 2017;**292**(11):169

[28] Perevalova NP, Sankov VA, Astafyeva EI, Zhupityaeva AS. Threshold magnitude for ionospheric TEC response to earthquakes. Journal of Atmospheric and Solar-Terrestrial Physics. 2014;**108**:77-90

[29] Jayachandran PT, Langley RB, MacDougall JW, Mushini SC, Pokhotelov D, Hamza AM, et al. The Canadian high arctic ionospheric network (CHAIN). Radio Science. 2009; **44**:RS0A03. DOI: 10.1029/2008RS004046

[30] Bougard B, Sleewaegen JM, Spogli L, Veettil SV, Monico JF. CIGALA: Challenging the solar maximum in Brazil with PolaRxS. In: Proceedings of the 24th International Technical Meeting of the Satellite Division of the Institute of Navigation 2011, ION GNSS. Vol. 2011. 2011. pp. 2572-2579

Chapter 2

The Influence of the Lower Ionospheric Disturbances on the Operating Conditions of Navigation Satellite Systems

Boris Gavrilov, Yuriy Poklad and Iliya Ryakhovskiy

Abstract

The study of the impact of ionospheric disturbances on the conditions of functioning of satellite communication and navigation systems and the development of methods to reduce this effect requires the development of methods for evaluating the parameters of ionospheric disturbances and their spatial and temporal distribution. Studies show that electron concentration disturbances, which can have a significant impact on the functioning of transionospheric radio channels, can occur both in the upper and lower ionosphere. At the same time, the methods of studying the dynamics of ionospheric disturbances in the lower ionosphere are not enough developed, and the interrelation of the lower and upper perturbations of the ionosphere is insufficiently studied. The aim of the work is an experimental study of disturbances of the upper and lower ionosphere in order to clarify the mechanisms of their relationship and study the spatiotemporal distribution of mid-latitude disturbances. The results obtained show that the contribution of the electron density disturbances in the D region to the total electron content of the ionosphere can be significant and considerably depends on the type of heliogeophysical processes.

Keywords: upper and lower ionosphere, traveling ionospheric disturbances, TEC, VLF signals, magnetic storms, solar X-ray flares

1. Introduction

The ionosphere is a region in the near-Earth space, where a number of technical systems vital for the life and safety of mankind (telecommunication, navigation, aircraft, surveillance systems, etc.) work continuously. These systems based on radio signals are sensitive to the varying electron density in the ionosphere. Its strong perturbations may cause failures and malfunction in these systems. So the investigation of the state and dynamics of the ionosphere and a prediction of irregularities and disturbances appearing are key questions.

Ionospheric disturbances are closely related to geomagnetic storms, solar flares, and other natural and anthropogenic processes [1–6]. The effect that the lower and upper ionospheres have on the propagation of a radio signal depends on their frequency. The F region is critical for the propagation of high-frequency (HF) waves. State and dynamics of the D and E regions define the conditions of propagation of low-frequency (LF) and very low-frequency (VLF) waves. Due to these reasons, HF and LF-VLF waves can be used to study F and D–E regions, respectively.

The impact of the processes of interaction in the lithosphere, atmosphere, ionosphere, and magnetosphere system on the upper and lower ionosphere and radio wave propagation was studied for decades [7–10]. Total electron content (TEC) values determined from data of dual-frequency measurements of global navigation satellite system (GNSS) signals are widely used to study the state and dynamics of the ionosphere [1, 3, 5, 6]. TEC is an integral of electron density in a tube with a cross section of 1 m^2 along the path of radio signal propagation from the navigation satellite to the receiver. It is assumed that TEC value mainly characterizes the state of the F region where (at least in quiet heliogeophysical conditions) the maximum electron density is observed.

Obtaining direct data on the state and dynamics of the lower ionosphere is a more complex experimental task, since at these altitudes the ionosondes, radars, and spacecraft practically do not work. The state of the lower ionosphere is often monitored by analyzing the characteristics of VLF (3–30 kHz) radio signals that propagate in the waveguide formed by the Earth's surface and the D region of the ionosphere. Variations in the amplitude and phase of VLF signals are mainly associated with changes in the state of the upper wall of the waveguide [11–13].

Despite the fact that both methods of ionosphere studying are quite effective, they are used separately as a rule, which does not allow investigating the relationship between the disturbances of the upper and lower ionosphere.

The focus of this article is an experimental study of the relationship between the perturbations of the upper and lower ionosphere.

The experiments were carried out during a strong geomagnetic storm and strong solar X-ray flares. Total electron content (TEC) data obtained from measurements of global navigation satellite system (GNSS) signals were used to study the F region. Information about the disturbances of the lower ionosphere is obtained by analyzing the amplitude and phase variations of VLF signals. Coordinated analysis of TEC and VLF signals is a powerful tool for studying interrelated processes in the D and F regions of the ionosphere. The results obtained strongly indicate their interconnected perturbations.

2. Experimental setup

Disturbances of the electron density and radio wave propagation in the D, E, and F regions of the ionosphere were investigated in the latitude range from 40° to 70° N and in the longitude range from 0° to 40° E. The "Mikhnevo" geophysical observatory (MIK, 54.9617° N, 37.7626° E) of the Institute of Geosphere Dynamics of the Russian Academy of Sciences (http://idg.chph.ras.ru/ru/watch/mikhnevo) continuously monitors the amplitudes and phases of signals in the frequency band from 9 to 30 kHz received from VLF stations located in Europe, Asia, and North America [14].

For the investigation of the upper ionosphere, we use the data of GPS receivers in Mikhnevo observatory and the worldwide GPS vertical TEC data included in the Madrigal database at MIT Haystack Observatory (http://www.openmadrigal.org/).

The Madrigal data contain TEC values with a time step of 5 min. These data were averaged over a 15-min interval and distributed over the 180° $\times$ 360° grid with a step of 1°.

The deviation of TEC from the median value is calculated by the formula

$$\Delta TEC(lat, long, t) = TEC(lat, long, t) - M, \tag{1}$$

where lat and long are latitude and longitude, t is UT, and M is the median value of TEC for the previous 27 days for the point with coordinates lat and long.

The equipment used to make the VLF signal measurements is the Metronix Analog/Digital Signal Conditioning Unit ADU-07 data logger connected to Metronix MFS-07 magnetic field sensor. MFS-07 is a high-frequency induction coil magnetometer, and two are mounted along the geographic north-south (X) and east-west (Y) axes. The magnetic field sensors cover a wide frequency range from 1 mHz up to 50 kHz and a dynamic range >130 dB and have excellent low noise characteristics (5×10^{-7} nT/$\sqrt{Hz}$ at 1000 Hz). ADU-07 unit and MFS-07 induction coils have a very stable transfer function over temperature and time. A GPS clock provides the timing signals.

Radio signals with a frequency below 30 kHz propagate in the Earth-ionosphere waveguide at distances of thousands and tens of thousands of kilometers with low attenuation. The relationship of the amplitude and phase of the VLF signal on the state of the D layer makes it possible to detect disturbances of the lower ionosphere in the path of the radio signal propagation. The transmitting stations were chosen so that the radio paths were under different azimuths over the territory of Europe. In our experiments the data of synchronous measurements and the signals from four transmitters were used: JXN (Gildeskål, Norway, 66.98° N, 13.87° E), GQD (Anthorn, UK, 54.91° N, 2.27° W), GBZ (Skelton, UK, 54.73° N, 2.88° W), and NAA (Cutler, USA, 44.65° N, 67.28° W).

When analyzing the parameters of the VLF signal, it should be taken into account that their variations are associated with changes in the parameters of the upper wall of the waveguide along the entire path of signal propagation. In order to localize disturbances in the lower ionosphere, data on the signal propagation along the path were additionally used, crossing the paths JXN-MIK, GQD-MIK, and NAA-MIK. We chose the data of Kiel Longwave Monitor (http://www.lf-radio.de/) where signals were received from Norwegian transmitter JXN (66.96° N, 13.90° E).

Radio signals received at the MIK and Kiel (54.4° N, 10.1° E) observatories were compared with data on variations in TEC of the ionosphere according to the Scripps Orbit and Permanent Array Center (SOPAC) http://sopacold.ucsd.edu/dataBrowser.shtml and Madrigal (http://www.openmadrigal.org) databases. Since we were

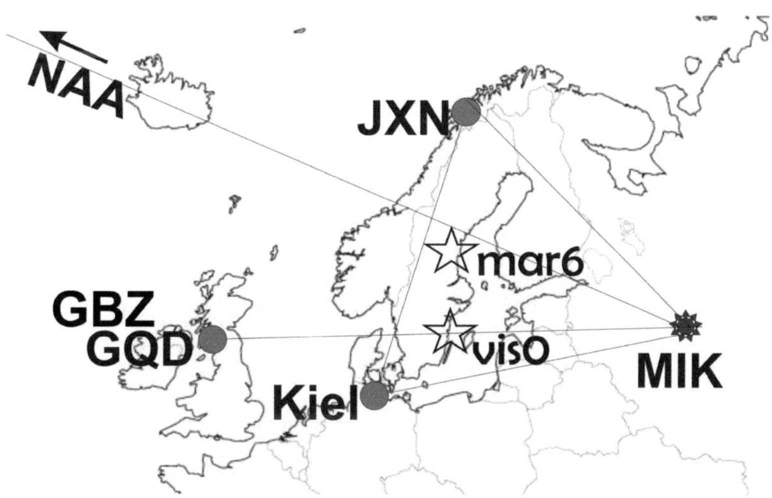

Figure 1.
The location of VLF transmitters (gray circles), GPS receivers (asterisks), and MIK observatory.

interested in the interrelated perturbations of the upper and lower ionospheres, GPS stations located near the used VLF signal traces were chosen. The location of the transmitters and measuring stations is shown in **Figure 1**.

The possibility of detection of interconnected disturbances in the upper and lower ionosphere by GPS and VLF receivers is shown by the example of the study of the ionospheric effects of the magnetic storm on 17 March 2015 and the solar X-ray flash on 6 September 2017. These events were chosen because they caused significant changes in the ionosphere, but the mechanisms of generation and evolution of ionospheric inhomogeneities during magnetic storms and solar X-ray flares are very different, which should be manifested in the pattern of the reaction of the D and F layers of the ionosphere to these phenomena.

3. St. Patrick's Day geomagnetic storm

The storm began on 15 March 2015 as a series of mid-level solar flares culminating in a class C9 flare at 02:13 UT. At 04:05 UT on 17 March 2015, the Advanced Composition Explorer (ACE) satellite recorded a sharp increase in the solar wind speed up to 500 km/s. The lowest value of the disturbance storm time index Dst exceeded 200 nT, the auroral activity index AE exceeded 2200, and the planetary index of the geomagnetic activity Kp reached a value of 8. Such values of these indices make it possible to define the event on 17 March 2015 as an extreme magnetic storm, which caused a storm in the ionosphere.

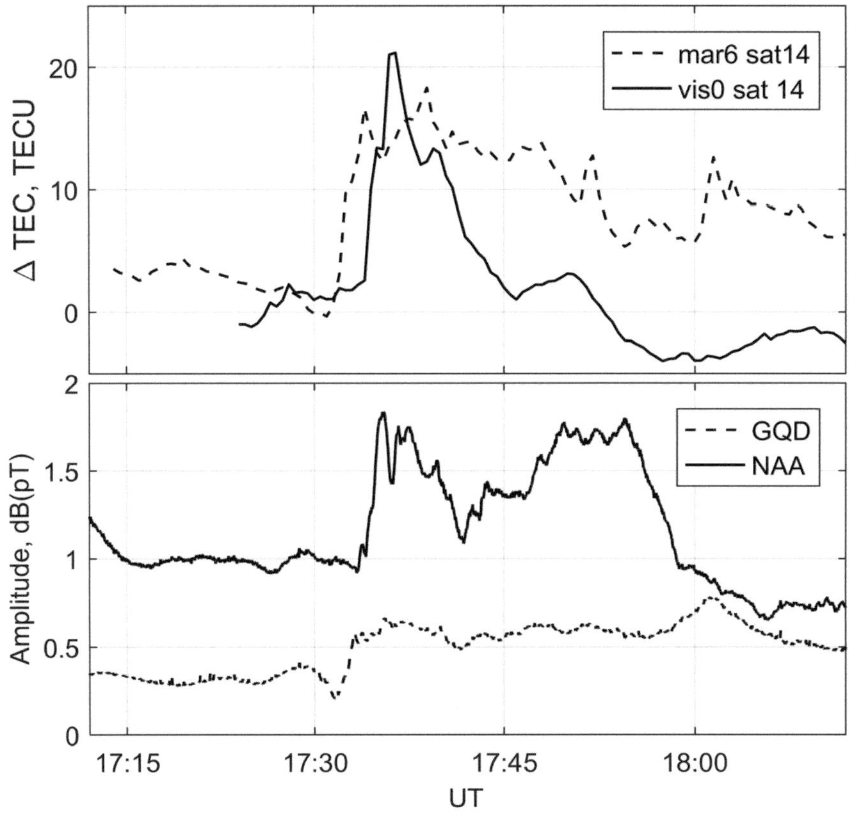

Figure 2.
Variations of TEC according to observatories mar6 and viso (top panel), VLF signal amplitude on NAA and GQD-MIK paths (bottom panel) on 17 March 2015.

The strongest geomagnetic storm in the current solar cycle has been studied in sufficient detail. Ground-based and space-born measurements demonstrate the response of the ionosphere to the geomagnetic storm. Astafyeva et al. and Borries et al. [15, 16] presented the results of investigation of the effects of the St. Patrick's Day ionospheric storm from the data of ground-based the GPS receivers,

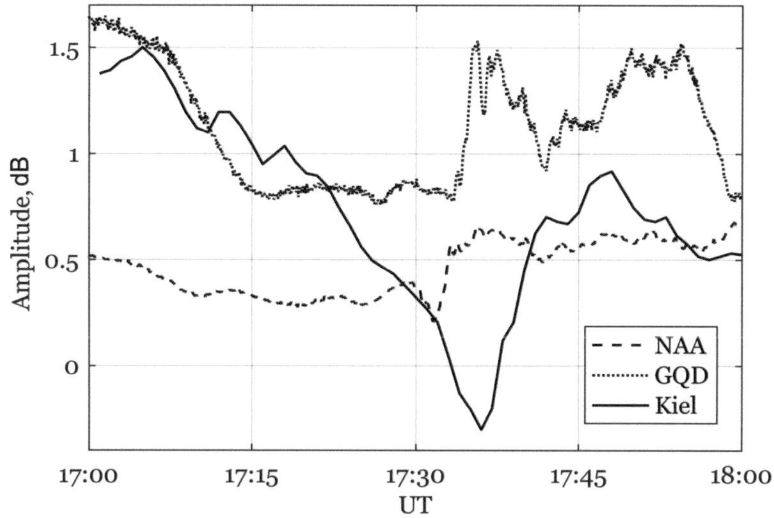

Figure 3.
Variations of VLF signal amplitude on GQD, NAA-MIK paths and on JXN-Kiel path (http://www.lf-radio.de/) from 17 to 18 UT on 17 March 2015.

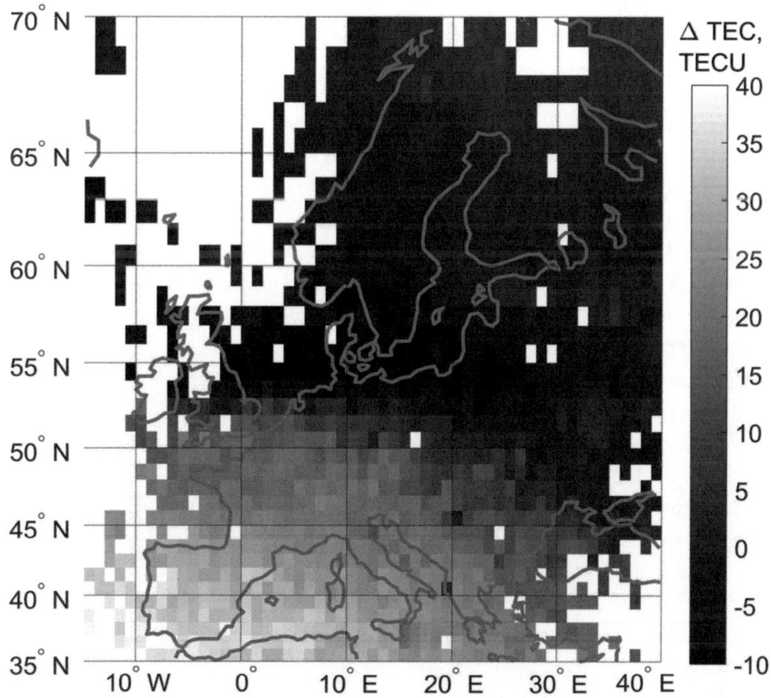

Figure 4.
The ΔTEC value determined by Eq. (1) for 17.5 UT on 17 March 2015.

ionosondes, and satellite missions. They reveal both a positive effect (TEC increase) at low- and mid-latitudes and positive and negative phases throughout all the latitudes. So, the results of these studies are mostly related to disturbances in the F region of the ionosphere. The effect of the magnetic storms in the lower ionosphere is less known due to the limited possibilities of ionosondes and incoherent scattering radars for the investigation of this region. Our data should allow to compare the results of the study of the F layer of the ionosphere with the effects observed in the lower ionosphere.

The upper panel of **Figure 2** shows the ionospheric variations of TEC calculated from data of GPS receivers located at stations vis0 and mar6. The lower panel shows the VLF signal amplitude variations on the NAA-MIK and GQD-MIK paths. TEC disturbances and VLF signal variations correspond to the main phase of the magnetic storm.

The **Figure 3** shows the change in the VLF signal amplitude on the JXN-Kiel path together with signals on the GQD-MIK and NAA-MIK paths. The maximum amplitude of all signals also corresponds to the main phase of the storm.

Figure 4 shows the distribution of TEC deviation from the previous 27 days over Europe at 17:30 UT according to data of the Madrigal network. The Madrigal data contain TEC values with a time step of 5 min. These data were averaged over a 15-min interval and distributed over the geographic grid with a step of 1°.

It can be seen that the strongest ionospheric perturbations are localized at this time in the region of our measurements around the Kiel and along the GQD-Mikhnevo path.

4. Ionospheric effects of the solar X-ray flares in September 2017

The solar X-ray flares were chosen as another high-energy event, different from the magnetic storm by the mechanisms of influence on the ionosphere. The main perturbation agent of the ionosphere is X-ray and ultraviolet radiation.

Monitoring of VLF signals is conducted in the Mikhnevo since 2014. The most powerful solar X-ray flares for this period occurred in early September 2017. Two solar X-ray flares X2.2 at 09 UT and X9.3 at 12 UT were observed on 6 September 2017. 10 September 2017 was observed X8.3 flare. But at this time, our receivers in Mikhnevo and part of VLF paths were in the region of evening terminator. So its flare was not used in our analysis.

To analyze flare effects in the upper and lower ionosphere, we used the same GPS stations and the same VLF signal paths that were used at observing the effects of the magnetic storm on 17 March 2015. Note that all measuring points and radio paths were located in the territory illuminated by the flashes.

Graphs of vertical TEC variations at flare 6 September 2017 at 12 UT according to GPS receiver located at mar6 and vis0 stations are shown in **Figure 5a**. Variations in the amplitude and phase of the radio signal received in the Mikhnevo from two VLF radio transmitters (GQD and NAA) are shown in **Figure 5b** and **c**. The maximum response to the flare was observed at 12 UT as a simultaneous jump in the ΔTEC and in phase and amplitude of VLF signals.

Comparing **Figures 2** and **5**, we can conclude that the growth of TEC during the X-ray flash was about 10 times less and the increase in the amplitude of the VLF signal was about 5 times greater than during the magnetic storm. To evaluate the effect of the X-ray flash on the additional ionization of the D layer, it is necessary to use theoretical models that allow to relate the parameters of the VLF signals to the change in the parameters of the lower ionosphere.

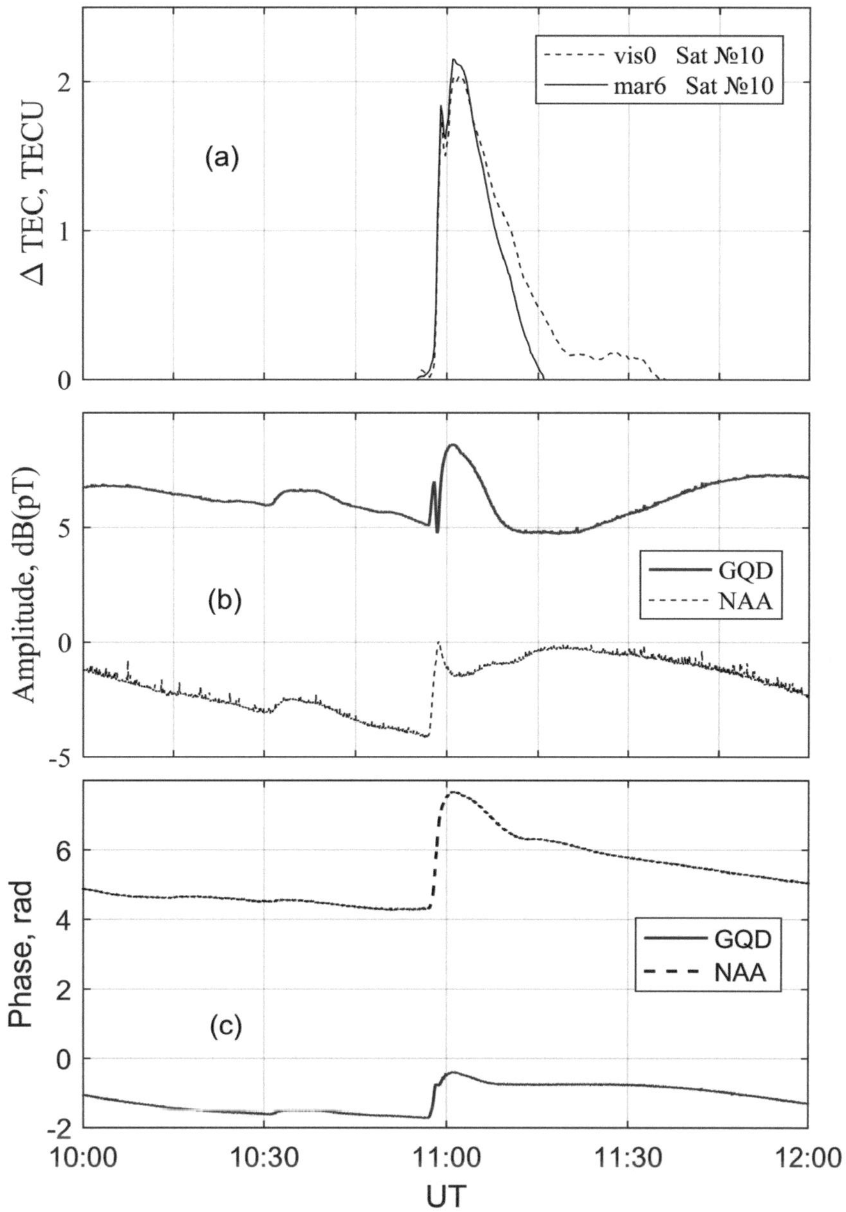

Figure 5.
(a) Variations of TEC according to GPS receiver mar6, (b) amplitude, and (c) phase of VLF signals on the paths JXN-MIK, GQD-MIK, and NAA-MIK on 6 September 2017.

One of the most common ways to describe the D region is the Ferguson-White model [17, 18]. According to this model, the altitude profile of the electron concentration in the lower ionosphere is described by the equation:

$$Ne(z) = 1.49 \cdot 10^7 \cdot \exp\left((\beta - 0.15)(z - h')\right) \cdot \exp\left(-0.15h'\right) \quad \left[cm^{-3}\right] \tag{2}$$

where h' is the referenced altitude of the ionosphere, β is the slope factor or sharpness of the electron concentration profile, and h is the current height.

The approach proposed in [19] can be used to estimate variations of these ionospheric parameters during solar flares. In this work, the effects of solar energetic phenomena on the lower ionosphere using parameters of subionospherically propagating VLF signals were studied. This is done in two steps. At the first stage, initial values of h' and β are selected for this VLF signal propagation path from empirical models [20–22] that take into account the impact on the value of h' and β zenith angle of the sun, latitude, day of the year, and solar activity in the form of the Wolf number. In the second stage, the standard for estimation of the VLF signals propagation Long Wavelength Propagation Capability (LWPC) code is used to estimate the amplitude change in the exponential ionosphere. As a result, the values of h' and β as a function of the X-ray flux were obtained.

The disadvantage of this approach is that the actual state of the ionosphere before a flash may differ from the model due to disturbances from such effects as magnetic storms and an increased X-ray flux.

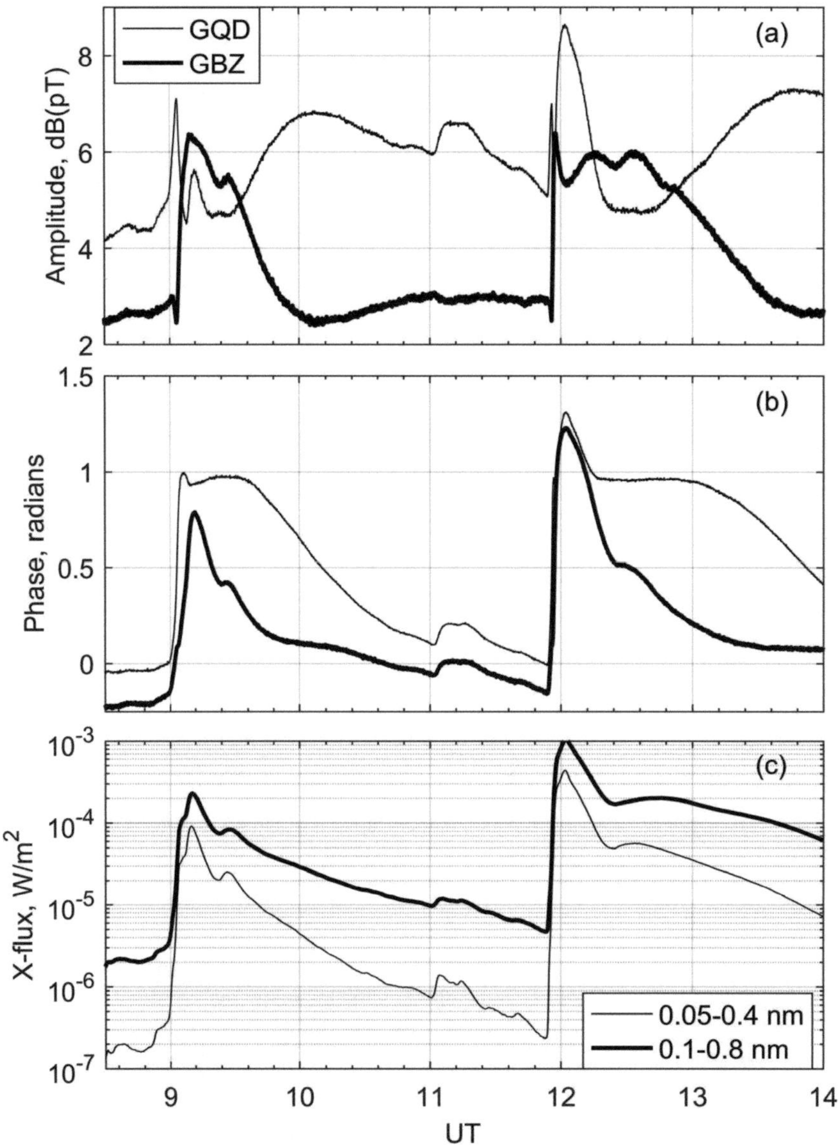

Figure 6.
Variations of the amplitude A^* (a) and phase P^* (b) of the signals from GQD and GBZ transmitters and X-ray flux (c) from GOES 15 satellite data (https://www.polarlicht-vorhersage.de/goes/2017-09-06_110000_ 2017-09-06_130000.png) on 6 September 2017.

In this paper, a different approach is used to evaluate changes in the lower ionosphere caused by X-ray flashes. The key difference between our method and [19] is the approach to determining the parameters of the ionosphere. We obtain initial conditions by processing experimental data on the amplitude and phase of VLF signals under the action of X-rays. We have developed a method for restoring the altitude profile of electron concentration in the D region of the ionosphere by using the amplitude and phase characteristics of signals from VLF transmitters on a two-frequency path. To implement this technique, the signals of two VLF transmitters located at a distance of 32 km from each other were used. GQD and GBZ transmitters operate at frequencies of 22.1 and 19.58 kHz, respectively. Taking into account the length of the path of about 2600 km, we can assume that the signals from these two stations are distributed along one two-frequency path.

The key difference between our method and [19] is the approach to determine the parameters of the undisturbed ionosphere. Statistical data do not take into account the impact on the ionosphere of factors not described by empirical models. We obtain initial conditions by processing experimental data on the amplitude and phase of VLF signals under the action of X-rays.

Let us consider in detail this technique on the example of ionosphere parameter recovery during X-ray flash of class X9.3 on 6 September 2017.

Figure 7 shows the experimental data on the variations of the amplitude and phase of the signals from GQD and GBZ stations. The bottom panel shows the X

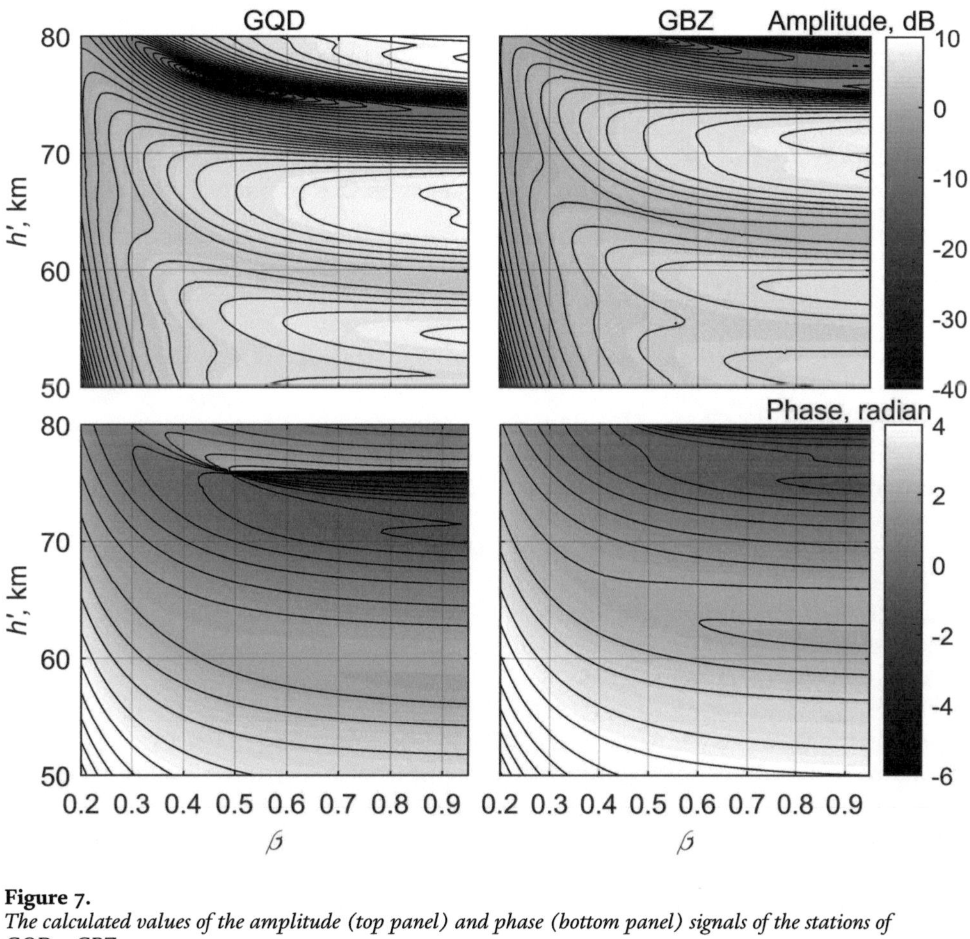

Figure 7.
The calculated values of the amplitude (top panel) and phase (bottom panel) signals of the stations of GQD u GBZ.

flux in two spectral bands according to the Geostationary Operational Environmental Satellite (GOES) data.

Denote the X9.3 flash start time as t_0 = 11:52:20 UT and the time of maximum radiation as t_{max} = 12:02:14 UT. Let variations of the amplitude and phase $dA_i(t)$ and $dP_i(t)$ be determined as

$$\begin{aligned} dA_i(t) &= A_i^*(t) - A_i^*(t_0) \\ dP_i(t) &= P_i^*(t) - P_i^*(t_0) \end{aligned} \tag{3}$$

where A_i^* and P_i^* are the measured amplitudes and phases of the signals of the GQD (i = 1) and GBZ (i = 2) transmitters shown in **Figure 6**.

On the other hand, the amplitude and phase of the signals of the GQD and GBZ transmitters depend on the parameters of the ionosphere h' and β on the signal propagation path. Suppose that for the whole GQD-/GBZ-Mikhnevo path, the parameters h' and β are the same. Then, the amplitude and phase values of the signals from the transmitters for all possible pairs of h' and β values were calculated using LWPC code. This code allows to calculate amplitude and phase of the VLF signal for the given path and the values of h' and β parameters. The calculations were carried out in the range of 50–90 km with 0.035 km increments for h' and 0.2–0.95 km^{-1} with 0.001 km^{-1} increments for β. The ranges of values of h' and β were selected according to [20–22]. Thus, four matrices with the size of 1143 × 751 elements with values of amplitudes and phases of VLF signals versus h' and β parameters were obtained. Let us denote the calculated values of the amplitude and phase as $A_i(h', \beta)$ and $P_i(h', \beta)$, where i = 1 for GQD and i = 2 for GBZ transmitters. The graphical representation of these data is shown in **Figure 7**.

Let us denote the parameters of the ionosphere at a time t_0 as h'_0 and β_0 and at a time t_{max} as h'_{max} and β_{max}. In the matrices $A_i(h', \beta)$ и $P_i(h', \beta)$, we can find all pairs of points (h_0', β_0) and (h'_{max}, β_{max}) for which the difference between the values of amplitudes and phases for t_0 and t_{max} coincides with that measured with given precision:

$$\begin{aligned} \left| A_i\left(h'_{max}, \beta_{max}\right) - A_i(h'_0, \beta_0) - dA_i(t_{max}) \right| &< \delta A \\ \left| P_i\left(h'_{max}, \beta_{max}\right) - P_i(h'_0, \beta_0) - dP_i(t_{max}) \right| &< \delta P \end{aligned} . \tag{4}$$

where δA = 0.12 dB and δP = 0.06 rad are the accuracy of estimation of the parameters of the ionosphere.

Based on the data [20–22], we assume that the parameters of the ionosphere at t_0 lie in the range of 68 < h_0' < 77 km and 0.22 < β_0 < 0.35 and the parameters of the ionosphere at the t_{max} lie in the range of 54 < h_{max}' < 68 and 0.31 < β_{max} < 0.95.

The regions of existence of pairs of points (h_0', β_0) and (h_{max}', β_{max}) that satisfy condition Eq. (4) are shown in **Figure 8** by extended gray areas. They show the entire possible range of ionospheric parameters before the flash Eq. (1) and at maximum X-ray radiation Eq. (2).

The range of the obtained parameter values is quite wide. Let us try to narrow down the ranges of values of h_0' and β_0. To do this, according to Eq. (3), we calculate the variations of the amplitude and phase of the signals for a time step of 15 seconds. So, for time from t_0 to t_{max}, we obtained 39 intermediate values of the variations of amplitudes and phases of the signals from the transmitters. Among the family of points (h_0', β_0), we find those for which there are such points (h_t', β_t) and for which the variation of amplitude and phase corresponds to that measured for all registered intermediate values:

$$\begin{vmatrix} A_i\left(h_t^{'},\beta_t\right) - A_i(h_0^{'},\beta_0) - dA_i(t) \end{vmatrix} < \delta A \\ \begin{vmatrix} P_i\left(h_t^{'},\beta_t\right) - P_i(h_0^{'},\beta_0) - dP_i(t) \end{vmatrix} < \delta P, \quad (5)$$

where $t_0 < t < t_{max}$. This family of values is shown in **Figure 8** as black points.

From **Figure 8** it is seen that the initial parameters of the ionosphere lie in a very narrow range. It is approximately 0.6 km in h' and 0.015 km^{-1} in β. The dispersion of the ionospheric parameters at the time of maximum flare is approximately the same in h' and 10 times larger in β (about 0.15). This is due to the fact that the flash was sufficiently powerful and at the time of the maximum X-ray flux, the "rigidity" of the upper wall of the waveguide became so large that its further increase had virtually ceased to influence the amplitude-phase characteristics of the received signals.

Further, for the initial parameters of the ionosphere that we calculated (i.e., $h' = 70.7$ and $\beta = 0.33$ km^{-1}), we can use Eq. (5) to continue the calculation of the parameters of the ionosphere for another time: during the decay phase of the X-ray flux, i.e., for time $t > t_{max}$, and for the time before flash, i.e., $t < t_0$. Thus, we can restore the time history of the parameters h' and β for the X-ray flash. These results are shown in **Figure 9b**. This figure clearly shows the advantage of our method. It can be seen how the parameters of the ionosphere changed before the flash from 11:30 to 11:55 UT. The parameter h' changed from 69.5 to 71 km and the β parameter from 0.36 to 0.33 km^{-1}. This is due to the relaxation of the ionosphere after the previous flare of class X2.2 that occurred at 9:00 UT. The same method was used for

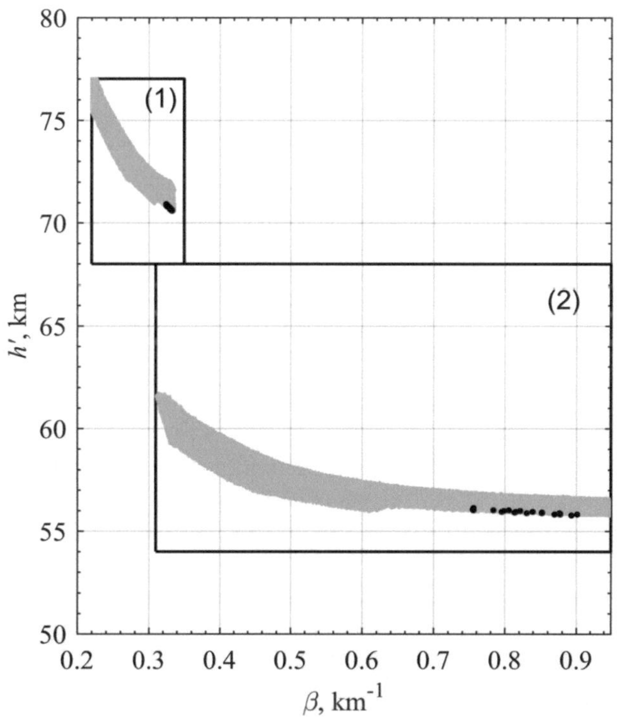

Figure 8.
The regions of ionospheric parameters h' and β before the flash Eq. (1) and at maximum X-ray radiation Eq. (2). The regions of existence of pairs of points (h_o', β_o) and (h_{max}', β_{max}) that satisfy the condition Eq. (4) are extended gray areas.

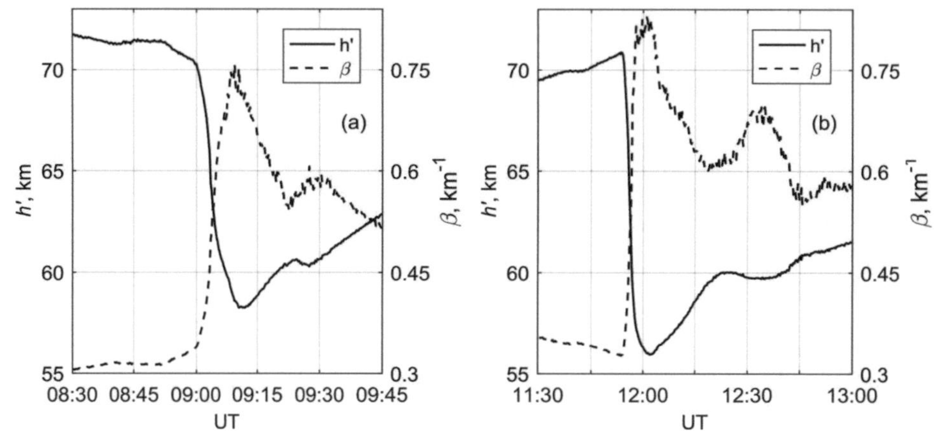

Figure 9.
The time history of the lower ionosphere parameters h′ and β for the X2.2 flash (a) and for the X9.3 flash (b) on 6 September 2017.

X2.2 flash at 9:00 UT on 6 September 2017. The results of these calculations are shown in the **Figure 9a**.

Knowing the parameters h' and β of the altitude profile of the electron concentration in the D region, it is possible to calculate the electronic content in it. Let us call this parameter DEC. To do this, the value of the electron concentration, that is, defined by Eq. (2), must be integrated in height in according to

$$DEC = \int_{50}^{h'+\delta h} 1.49 \cdot 10^7 \cdot e^{(\beta-0.15)(z-h')} e^{-0.15h'} dz \tag{6}$$

The lower limit of integration is not very important, since according to Eq. (2) for small z, the electron concentration of N_e is minimal. But there is a problem of choosing the upper limit of integration. This is due to the fact that the according to Eq. (2), the ionosphere shows the exponential growth of the electron concentration with height. At the same time, the electromagnetic wave cannot penetrate into the region of high concentrations and, therefore, does not carry any information about the state of the ionosphere at these heights. So, the VLF radio waves should be reflected in a layer of thickness of the order of wavelength. Therefore, it would be reasonable to carry out the integration up to the height $h' + \delta h$, where δh is a value comparable to the wavelength.

Now consider the effect of solar X-ray flares in the upper and lower ionosphere comparing the change in TEC of the ionosphere according to GNSS receiver data to electronic content in the D region (let us call this parameter DEC), according to the parameters of VLF radio signals.

Figure 10 shows DEC and TEC variations caused by flashes X2.2 (a) and X9.3 (b) on 6 September 2017. Here, to calculate DEC the integration of the electron density N_e was carried out up to a height of h' + 12 km. TEC was calculated from the GPS receiver data installed in the Mikhnevo observatory.

You can see that at the solar flash X2.2, (a) the amplitude of the increase in TEC and DEC was about four times less than during the flash X9.3. So, the change of the amplitude of the perturbations of both the upper and lower ionosphere was directly proportional to the change in the X-ray flux. But the most interesting result of these measurements is the proximity of the electron density perturbations in the upper and lower ionosphere. This result shows that the increase in TEC in both cases was provided by an increase in the electron concentration in the lower ionosphere.

Figure 11 shows the deviations of TEC from the median value over the preceding month of 6 September 2017. This map was constructed from the Madrigal navigation network by using algorithm Eq. (1) as in the previous section. Comparison of **Figures 4** and **11** shows that the maximum deviation of TEC during a solar flare is about four times less than in the main phase of a strong magnetic storm.

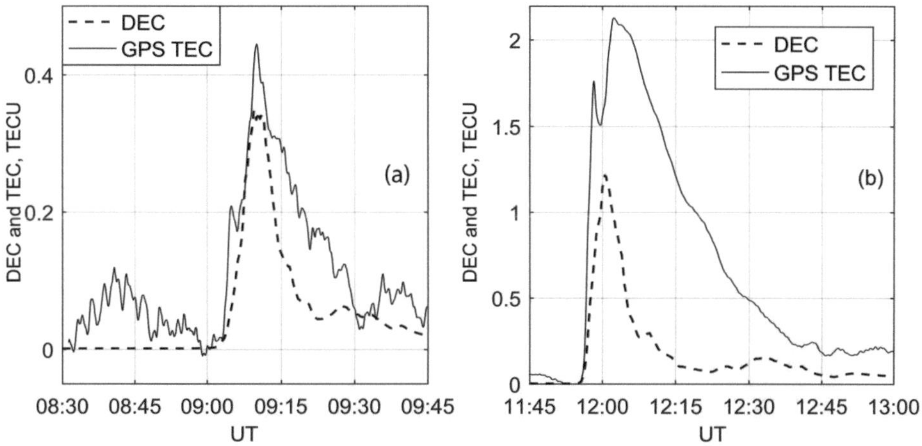

Figure 10.
Variations of DEC and TEC during solar flashes X2.2 (a) and X9.3 (b) on 6 September 2017.

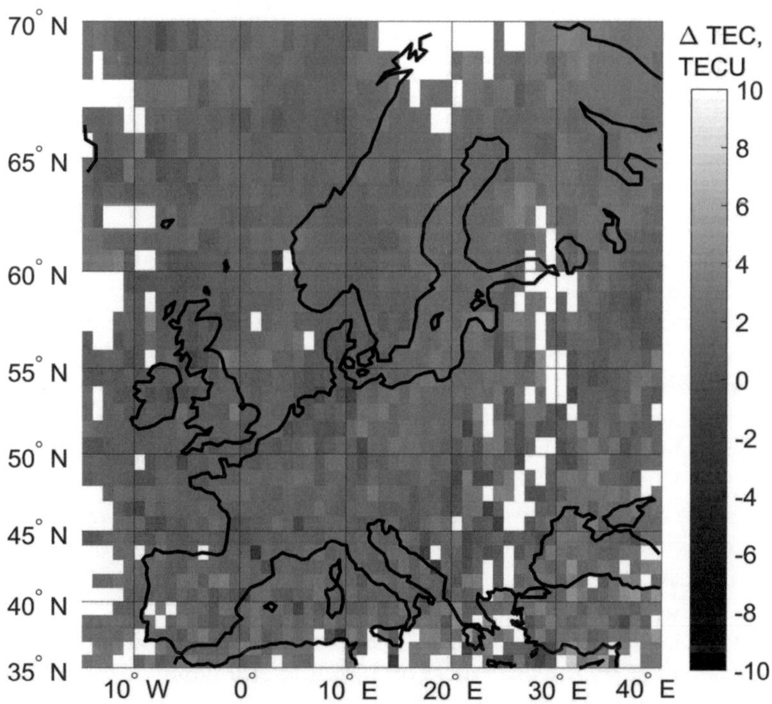

Figure 11.
The ΔTEC value determined by Eq. (1) for 6 September 2017.

5. Conclusion

The study of the influence of heliogeophysical phenomena on the conditions of GNSS functioning was carried out under strong magnetic storms and powerful solar X-ray flares. Experiments on these types of events were carried out in one latitudinal zone using one set of data sources. This allows not only to study the disturbances of the upper and lower ionospheres for different geophysical processes but also to compare their nature and magnitude.

The combined analysis of data from GPS receivers, data from the Madrigal network, and data from measurements of VLF radio signals provides a fairly complete picture of the effects caused by X-ray flashes and magnetic super storm in the upper and lower ionospheres.

Both during the solar X-ray flares on 6 September 2017 and during the magnetic storm on 17 March 2015, the significant changes in the electron concentration of the ionosphere were observed. But the amplitude of these changes and most importantly the ratio of the growth of the electron concentration in the D and F regions of the ionosphere differed significantly in these two events.

During the geomagnetic storm on 17 March 2015, TEC disturbances and VLF signal variations correspond to its main phase. TEC increase was about 15–20 TECU (**Figure 2a**). The perturbation of the F layer at solar X-ray flares was much lower and did not exceed 2.5 TECU (**Figure 5a**), so it was approximately eight times less than during the magnetic storm.

Comparison of TEC estimated from the Madrigal GPS network (**Figures 4** and **10**) with the data obtained for individual stations is not correct. However, it can be seen that according to Madrigal, TEC increase during the magnetic storm was significantly stronger. It is also obvious that during the solar flashes, there is a much more uniform distribution of TEC, as it should be in the conditions of illumination by the radiation of the flash throughout Europe.

The analysis of the results of VLF radio signal amplitude variations on the paths from several European and American VLF transmitters to the receiver at Mikhnevo observatory during St. Patrick's Day magnetic storm on 17 March 2015 and solar X-ray flares on 6 September 2017 shows the effects of the both events to the lower ionosphere as well.

To estimate the lower ionospheric contribution to the TEC value, we have developed a method for restoring the high-altitude profile of electron concentration in the D region of the ionosphere by using the amplitude and phase characteristics of signals from VLF transmitters on a two-frequency path.

This method is based on known approaches to solving this problem [17–19]. But if in these works the determination of the parameters of the lower ionosphere was carried out by analyzing the contribution to the ionization of X-ray radiation and as the initial (before the flash) parameters of the ionosphere were used parameters determined by models, our technique allows to calculate parameters of the lower ionosphere, in which it was before the disturbance using only the measurement data.

A comparison of TEC value calculated from GPS receiver data with the calculation of the electron content in the D region from the data of the VLF radio signal parameters indicates the possibility of a significant contribution of the lower ionosphere, at least during the powerful X-ray flashes.

We have not been able to calculate DEC for magnetic storm conditions. Changing the parameters of VLF signals was complex and ambiguous. However, the fact that the increase in the amplitudes of VLF signals during the storm was close to the growth of this value during solar X-ray flares and the growth of TEC in a storm significantly exceeded the growth of TEC during solar X-ray flares clearly indicates a substantially smaller contribution of DEC to TEC in a magnetic storm than in solar X-ray flares.

So, the combined analysis of variations of GNSS signals and signals from VLF radio stations is an effective method for studying the interrelated processes in the upper and lower ionosphere [23, 24].

We believe that an important result of our research is to demonstrate that the interpretation of data on the total electronic content of the ionosphere obtained from the data of GNSS receivers should take into account the contribution that can make to the TEC value by the change of electron concentration in the lower ionosphere. On the other hand, the occurrence of significant electron density perturbations in the lower ionosphere should be taken into account in the analysis of factors affecting the accuracy and reliability of GNSS operation.

Acknowledgements

This study was conducted within state research targets AAAA-A17-117112350014-8 and 0146-2015-0017.

The authors are sincerely grateful to Madrigal (http://www.openmadrigal.org/), UK Solar Data Centre (https://www.ukssdc.ac.uk/cgi-bin/digisondes/cost_database.pl), and Kiel Longwave Monitor (http://www.lf-radio.de/) for providing them with geophysical data and to Johns Hopkins University Applied Physics Laboratory for providing them with DMSP satellite data (http://ssusi.jhuapl.edu/).

Author details

Boris Gavrilov*, Yuriy Poklad and Iliya Ryakhovskiy
Institute of Geosphere Dynamics RAS, Moscow, Russian Federation

*Address all correspondence to: boris.gavrilov34@gmail.com

IntechOpen

© 2019 The Author(s). Licensee IntechOpen. This chapter is distributed under the terms of the Creative Commons Attribution License (http://creativecommons.org/licenses/by/3.0), which permits unrestricted use, distribution, and reproduction in any medium, provided the original work is properly cited.

References

[1] Afraimovich EL, Boitman ON, Zhovty EI, et al. Dynamics and anisotropy of traveling ionospheric disturbances as deduced from transionospheric sounding data. Radio Science. 1999;**34**(2):477-487. DOI: 10.1029/1998RS900004

[2] Demyanov VV, Yu V. Yasyukevich S, Jin S. Effects of solar radio emission and ionospheric irregularities on GPS/GLONASS performance. In: Jin S, editor. Geodetic Sciences – Observations, Modeling and Applications. InTech; 2013. pp. 177-222. ISBN: 978-953-51-1144-3. DOI: 10.5772/54568

[3] Demyanov VV, Zhang X, Lu X. Moderate geomagnetic storm condition, WAAS alerts and real GPS positioning quality. Journal of Atmospheric Science Research. 2019;**2**(1). DOI: 10.30564/jasr.v2i1.343

[4] Ryakhovskiy IA et al. Ionization of the lower ionosphere during the X-ray solar flare on September 6, 2017. In: Proceedings of SPIE 10833, 24th International Symposium on Atmospheric and Ocean Optics; 2018. p. 108339Y. DOI: 10.1117/12.2504402

[5] Demyanov VV, Yasyukevich YV. GNSS carrier phase noise as a promising means to reconstruct fine structure of the ionosphere. Journal of Aeronautics & Aerospace Engineering. 2018;**7**:60. DOI: 10.4172/2168-9792-C1-22

[6] Afraimovich EL, Astafyeva EI, Demyanov VV, Gamayunov IF. Mid-latitude amplitude scintillation of GPS signals and GPS performance slips. Advances in Space Research. 2009;**43**: 964-972. DOI: 10.1016/j.asr.2008.09.015

[7] Afraimovich EL, Ding F, Kiryushkin V, Astafyeva E, Jin S, San'kov V. TEC response to the 2008 Wenchuan earthquake in comparison with other strong earthquakes. International Journal of Remote Sensing. 2010;**31**(13):3601-3613. DOI: 10.1080/01431161003727747

[8] Hayakawa M, Molchanov OA. Seismo Electromagnetics: Lithosphere-Atmosphere-Ionosphere Coupling. Tokyo: Terra Sci.; 2008. p. 477

[9] Perevalova NP, Sankov VA, Astafyeva EI, Zhupityaeva AS. Threshold magnitude for ionospheric TEC response to earthquakes. Journal of Atmospheric and Solar-Terrestrial Physics. 2014;**108**:77-90

[10] Kumar A, Kumar S. Space weather effects on the low latitude D-region ionosphere during solar minimum. Earth, Planets and Space. 2014. p. 76. DOI: 10.1186/1880-5981-66-76

[11] Gavrilov BG, Zetser Yu I, Ryakhovskii IA, Poklad Yu V, Ermak VM. Remote sensing of ELF/VLF radiation induced in experiments on artificial modification of the ionosphere. Geomagnetism and Aeronomy. 2015; **5**(4):450-456. DOI: 10.7868/S0016794015040045

[12] Han F, Cummer SA, Li J, Lu G. Daytime ionospheric D region sharpness derived from VLF radio atmospherics. Journal of Geophysical Research. 2011; **116**(5). DOI: 10.1029/2010JA016299

[13] Maurya AK, Veenadhari B, Singh R, et al. Nighttime D region electron density measurements from ELF-VLF tweek radio atmospherics recorded at low latitudes. Journal of Geophysical Research. 2012;**117**(A11). DOI: 10.1029/202JA017876

[14] Gavrilov BG, Zetser Yu I, Lyakhov AN, Poklad Yu V, Ryakhovskii IA. Spatiotemporal distributions of the electron density in the ionosphere by records of the total electron content and phase of VLF radio

signals. Geomagnetism and Aeronomy. 2017;**57**(4):461-470. DOI: 10.7868/S001679401704006X

[15] Astafyeva E, Zakharenkova IM, Forste M. Ionospheric response to the 2015 St. Patrick's Day storm: A global multi-instrumental overview. Journal of Geophysical Research, Space Physics. 2015;**120**(10):9023-9037. DOI: 10.1002/2015JA021629

[16] Borries C, Mahrous AM, Ellahouny NM, Badeke R. Multiple ionospheric perturbations during the Saint Patrick's Day storm 2015 in the European—African sector. Journal of Geophysical Research, Space Physics. 2016;**121**(11):11333-11345. DOI: 10.1002/2016JA023178

[17] Ferguson J, Snyder FP. Computer programs for assessment of long wavelength radio communications. Tech. Doc. 1773, DTIC AD-B144 839. Alexandria, VA: Naval Ocean System Center, Defense Technical Information Center; 1990

[18] Wait JR, Spies KP. Characteristics of the Earth—Ionosphere waveguide for VLF radio waves. Natl. Bur. of Stand. Technical Note; Boulder, Colo; 1964. p. 300

[19] Basak T. Study of the effects on lower ionosphere due to solar phenomena using very low frequency radio wave propagation [thesis]. Department of Physics, University of Calcutta; 2013

[20] Thomson NR. Experimental daytime VLF ionospheric parameters. Journal of Atmospheric and Terrestrial Physics. 1993;**55**(2):173-184. DOI: 10.1016/0021-9169(93)90122

[21] Davis RM, Berry LA. A revised model of the electron density in the lower ionosphere. Tech. Rept TR I I l-77. Defense Commun. Agency Command Control Tech. Center, Washington, DC. NTIS, AD 17883. 1997. p. 58

[22] Gambill B. Normal D-region models for weapon effects code. Defence Nuclear Agency Report. DNA 62715H; 1985

[23] Gavrilov BG, Zetser YI, Lyakhov AN, Poklad YV, Ryakhovskii IA. Correlated disturbances of the upper and lower ionosphere from synchronous measurements of parameters of GNSS signals and VLF radio signals. Cosmic Research. 2019;**57**(1):36-43. DOI: 10.1134/S0010952519010039

[24] Lyakhov AN, Korsunskaya JA, Poklad YV, et al. The numerical simulation of the 2017 september solar X-flares impact on the midlatitude lower ionosphere. In: Proceedings of SPIE 10833, 24th International Symposium on Atmospheric and Ocean Optics; 108339M; 2018. DOI: 10.1117/12.2504292

Chapter 3

Real-Time Monitoring of Ionospheric Irregularities and TEC Perturbations

Giorgio Savastano and Michela Ravanelli

Abstract

The ionosphere is a part of the upper atmosphere that is a threat to GNSS and satellite telecommunication systems. In this chapter, we will dive into the GNSS real-time monitoring of ionospheric irregularities and TEC perturbations, with a focus on the detection of small- and medium-scale traveling ionospheric disturbances (TIDs) for natural hazard applications. We will describe the Variometric Approach for Real-Time Ionosphere Observation (VARION) algorithm, which is capable of estimating TEC variations in real time, and it was used to detect tsunami-induced TIDs. In particular, the analytical and physical implications of applying the VARION algorithm both to GNSS dual-frequency MEO (medium Earth orbit) and GEO (geostationary orbit) satellites will be provided, thus highlighting its relevance for natural hazard early warning systems and real-time monitoring of ionospheric irregularities.

Keywords: VARION algorithm, GNSS, GEO, traveling ionospheric disturbances, tsunami early warning systems, ionospheric irregularities

1. Introduction

As the title of this book suggests, the Earth's atmosphere represents a threat for GNSS and telecommunications satellites. In particular, the charged component of the upper atmosphere, the ionosphere, is responsible for errors in GNSS positioning that can reach values of tens of meters for single-frequency GNSS receivers [1, 2]. These errors have to be corrected or eliminated in order to make GNSS a valuable scientific instrument for geodesy and geodynamics applications.

However, the use of GNSS signals is nowdays not only limited to the estimation of the receiver's position, but it has eventually become a key instrument for ionospheric and tropospheric remote sensing studies and for soil features (GNSS reflectometry) [3]. In particular, GNSS can be used to monitor the ionosphere at different time and space scales. On a global scale, GNSS observations are used to generate global ionosphere maps (GIM) by interpolating in both space and time measurements of TEC from stations distributed around the world [4]. On a regional scale, the same signals can be used to detect fast ionospheric disturbances, such as TIDs with periods of minutes to about 1 h [5] and ionospheric scintillation with periods of seconds [6, 7].

IntechOpen

The ionosphere is a very important region of the atmosphere as it carries much valuable information about the Earth's system. In fact, the ionosphere is affected from both ends: (a) from above by space weather, such as geomagnetic storms induced by strong solar events, and (b) from below by events such as extreme terrestrial weather and natural hazards.

In this chapter, we focus on the real-time monitoring of ionospheric irregularities and TEC perturbations through the application of the VARION algorithm. In Section 2, we review the main mechanisms by which numerous near-ground geophysical (e.g., earthquakes, volcano eruptions, tsunamis) and man-made (e.g., rocket launches) events induce variations in electron density in the ionosphere. In Section 3, we describe the VARION algorithm, which is capable of estimating in real- time changes in the ionospheres' TEC using stand-alone GNSS receivers and can be used for real-time ionosphere remote sensing. In Section 4 we present the main results of the application of the VARION method for two case studies: the 2012 Haida Gwaii tsunami event and a Falcon 9 rocket launch. In Section 5 we present our conclusions.

2. Earth's surface and ionosphere coupling mechanisms

Acoustic and gravity waves are the two main mechanisms by which energy produced by geophysical events at the Earth's surface can propagate in the atmosphere [8]. The coupling of these atmospheric waves with the ionospheric electron density [9] produces deviations in TEC from the dominant diurnal variation. Traveling ionospheric disturbances (TIDs) are the ionospheric manifestation of these AGWs' induced TEC perturbations. In several applications, such as TID detection, the deviations (also known as fluctuations or perturbations) from the background level are of interest [10, 11]. Other mechanisms by which the ionospheric plasma highly deviates from the dominant diurnal variability are the chemical processes responsible for the ionospheric hole induced by rockets. These processes were described as the interactions between water (H_2O) and hydrogen (H_2) molecules in the exhaust plume and electrons in the ionosphere, through dissociative recombination.

2.1 Acoustic waves

Pressure-induced TEC anomalies from earthquakes were widely observed in the last decade, for example, coseismic ionospheric disturbances (CIDs) were documented with the 2003 M_W 8.3 Tokachi-Oki, Japan and the 2008 M_W 8.1 Wenchuan, China earthquakes [12] observed at Japanese GEONET sites. CIDs produced by the 2011 M_W 9.0 Tohoku-Oki, Japan earthquake were reported by several independent research groups [13, 14]. Volcanic eruptions can also excite acoustic waves and induce anomalies in the TEC measurements [15].

When an earthquake occurs, shock acoustic waves (SAWs) are produced in the proximity of the epicenter (within 500 km), and secondary acoustic waves are caused by surface Rayleigh waves propagating far from the epicenter. These pressure waves, upon reaching the ionosphere, will locally affect electron density through particle collisions between the neutral atmosphere and the ionospheric plasma [16]. SAWs, governed primarily by longitudinal compression, can propagate through the atmosphere at the sound speed which varies from several hundred m/s near sea level to 1 km/s at 400 km altitude [17]. At the height of the ionosphere F layer, it is about 800–1000 m/s [18], so it takes between 10 and 15 min to reach the ionosphere and cause the abovementioned disturbance (CID) [19]. Their waveform is "N-type wave," consisting of leading and trailing shocks connected by smooth linear transition regions. The waveform arises from nonlinear propagation effects: the amplitude of N waves depends on earthquake magnitude, losses of shock fronts, neutral wind speed, etc. This means that also CID is N-shaped and propagates at such velocity [18]. Rayleigh waves travel along the Earth surface at a velocity of 3–4 km/s. They propagate in the form of a train consisting of several oscillations whose typical period is about ten of seconds [20]. As already mentioned, they trigger secondary acoustic waves emitted in the form of the same train, propagating at sound speed. These waves also appear as CID 10–15 min after the earthquake.

It is important to highlight that only acoustic waves which have a frequency greater than the cutoff frequency can propagate up to the ionosphere [21]. Such frequency is defined as $\omega_a = \frac{\gamma g}{2c_S}$ where c_S is the speed of sound and γ and g are, respectively, the specific heat ratio (of the atmosphere) and gravitational acceleration [22, 23]. Thus, waves with a frequency greater than the cutoff one can reach the ionosphere; otherwise their amplitude decreases exponentially with altitude [22], and in this case, the waves are named evanescent. The typical values of cutoff frequency fall within the range 2.1–3.3 mHz [15, 22].

2.2 Gravity waves

Gravity waves (GW) form when air parcels are lifted due to particular fluid dynamic and then pulled down by buoyancy in an oscillating manner. This can occur when air passes over mountain chains [24] or when a "mountain," which is read as tsunami wave, moves with a certain velocity. Let us imagine the displacement of a volume of atmospheric air from its equilibrium position; it will then find itself surrounded by air with different density. Buoyant forces will try to bring the volume of air back to the undisturbed position, but these restoring forces will overshoot the target and lead it to oscillate about its neutral buoyancy altitude. It will continue this oscillation about an equilibrium point, generating a gravity wave that can propagate up through the ionosphere.

Perturbations at the surface that have periods longer than the time needed for the atmosphere to respond under the restoring force of buoyancy will successfully propagate upward. This is known as the Brunt-Vaisala frequency N and represents the maximum frequency for vertically propagating gravity waves. $N = \sqrt{(g/\theta)(d\theta/dz)}$ where g is the gravitational acceleration, θ is the potential temperature (the temperature that a parcel of air would attain if adiabatically brought to the ground), and z is the altitude.

Tsunamis have periods longer than this frequency and thus excite atmospheric gravity waves (AGWs) that can propagate upward in the atmosphere and ultimately cause perturbations in the ionospheric electron density. As the kinetic energy is conserved up to an altitude of about 200 km, and air density decreases exponentially with altitude, the AGWs are then strongly amplified in the atmosphere. The ratio of the amplitude of the velocity wave between the ionospheric height and the ground level is about 10^4–10^5 [25]. This fact was first established in Daniels [26] and was theoretically further developed in Hines [27, 28]. Therefore, it is possible to remotely detect the effects of ocean tsunamis by observing perturbations in the ionosphere. In detail, AGWs which have frequency lower than the Brunt-Vaisala frequency can propagate up through the ionosphere [22]. In the Earth's atmosphere, it depends on the altitude, and it varies from 3.3 to 1.1 mHz (typical value is 2.9 mHz [22]), corresponding to a buoyancy period of 5 min at sea level and about 15 min at 400 km altitude, near the F region peak of the ionosphere [19].

TIDs can be detected using different observing methods, including ionosondes [29]; ground-based GPS total electron content (TEC) [17, 30]; dual-frequency, space-based altimeters [31]; incoherent backscatter radar (ISR) [32]; and space-based GNSS-RO measurements [33]. Perturbations in the neutral atmosphere after the 2011 Tohoku-Oki tsunami event have also been detected using accelerometers and thruster data from the GOCE mission [34]. Several other causes are responsible for TIDs, such as intense or large-scale tropospheric weather [35], geomagnetic and auroral activity [36, 37], and earthquakes [38–40]. For this reason, the relationship between detected TIDs and those that are induced by a tsunami has to be proven, for example, by verifying that the horizontal speed, direction, and spectral bandwidth of the TIDs match that of the ocean tsunami [5].

The vertical propagation speed of an atmospheric gravity wave at these periods is 40–50 m/s [41], so these perturbations should first be observed about 2 h after the onset of the tsunami. The TEC anomalies can be identified by their horizontal propagation speed, which is much slower (200–300 m/s) than that of the acoustic TID or Rayleigh-wave-induced anomalies and follows the propagation speed of the tsunami itself, which is, much like the Rayleigh waves in the acoustic case, a moving source of gravity waves. However, following the 2011 M_W 9.0 Tohoku-Oki, Japan event, which provided dense near-field TEC observations, it was noted that the onset of the gravity-wave-induced TEC anomalies was shorter, at about 30 min after the start of the earthquake, and not the 1.5–2 h predicted by previous theoretical computations [17]. This is explained as evidence that it might not be necessary for the gravity wave to reach the F layer peak (around 300 km altitude) for the TEC disturbance to be measurable. Rather, disturbances at lower altitudes within the E layer and the lower portion of the F layer might be substantial enough to be seen in the TEC observations. This is supported by previous modeling results that showed significant TEC perturbations over a broad area around the F layer peak [14]. Through comparisons with tsunami simulations of the event, it was convincingly demonstrated that the tsunami itself must be the source of the observed gravity waves [17]. In light of these observations, ionospheric soundings may be used to monitor tsunamis and issue warnings in advance of their arrival at the coast [3, 5].

2.3 Traveling ionospheric disturbances

Disturbances in the ionosphere naturally occur at many different scales. On a planetary scale, Rossby waves result from latitudinal variations in the strength of the Coriolis effect and have wavelengths of 1000s of km, while, at smaller scales, acoustic gravity waves induced by natural hazards have typical wavelengths in the range of 10-300 km. Based on their phase velocity, wave period, and horizontal wavelength, TIDs are often classified into medium-scale TID (MSTID) and large-scale TID (LSTID). Some guidelines on the properties of these two groups are summarized in **Table 1**, which was created from [42, 43].

In this chapter, we mainly take into account MSTIDs, as they are the one typically generated by tsunami waves and other natural hazards.

	Period [min]	Phase velocity [m/s]	Horizontal wavelength [km]
Large scale	30–300	400–1000	1000–3000
Medium scale	10–60	50–300	10–500

Table 1.
TID classification based on phase velocity, wave period, and horizontal wavelength.

2.4 Dissociative recombination

Several studies were carried out to analyze the ionospheric responses to rocket launches. The first detection of a localized reduction of ionization due to the interaction between the ionosphere and the exhaust plume of the Vanguard II rocket was reported in [44]. More than a decade after that observation, a sudden decrease in total electron content (TEC) was observed after the 1973 NASA's Skylab launch [45] by measuring the Faraday rotation of radio signals from a geostationary satellite. This study [45] was reported a dramatic bite-out of more than 50% of the TEC magnitude having a duration of nearly 4 h and spatial extent of about 1000 km radius. The chemical processes responsible for the ionospheric hole were described as the interactions between water (H_2O) and hydrogen (H_2) molecules in the exhaust plume and electrons in the ionosphere, through dissociative recombination. At the level of concentration at which the reactants (H_2O and H_2) were added to the ionosphere by the rocket's engines, the loss process became 100 times more efficient than the normal loss mechanism in the ionosphere (e.g., N_2). Localized plasma density depletions during rocket launches were detected also using other measurement techniques, such as ground-based incoherent scatter radar and digisonde [46, 47] and continuous Global Positioning System (GPS) receivers [48, 49].

3. VARION approach

Multiple algorithms were developed to estimate useful ionospheric parameters from GNSS signals, such as absolute TEC measurements [4, 50], relative TEC [51, 52], and TEC variations [5]. In this section, we review the main concepts of the VARION approach, which was first presented in [5] for GNSS satellites (Section 3.1) and subsequently expanded to geostationary satellites in [53] (Section 3.2).

3.1 VARION-GNSS

The VARION approach is based on single time differences of geometry-free combinations of GNSS carrier-phase measurements ($L_1 - L_2$), using a stand-alone GNSS receiver and standard GNSS broadcast orbits available in real time. The unknown carrier-phase ambiguity can be considered constant between two consecutive epochs as long as no cycle slips occur. In the case that a cycle slip does occur, then the phase jump can be removed in real time as it represents an outlier in the time series analysis. The receiver and the satellite IFBs in the carrier-phase ionospheric observable are also assumed as constant for a given period [54]. Multipath terms cannot be considered constant between epochs for sampling rates greater than 1 second [55]. However, these terms can be mitigated by applying an elevation cutoff mask of 20 degrees or higher and will be ignored in the following equations for the sake of simplicity. For these reasons, we can write the geometry-free time single-difference observation equation [5], with no need of estimate in real time the phase ambiguity and the IFB:

$$L_{GF}(t+1) - L_{GF}(t) = \frac{f_1^2 - f_2^2}{f_2^2}\left[I_{1R}^{\ S}(t+1) - I_{1R}^{\ S}(t)\right] \tag{1}$$

where the term L_{GF} refers to the geometry-free combination and f_1 and f_2 are the two frequencies in L-band transmitted by any GNSS satellites. Taking into account the ionospheric refraction along the geometric range, we compute the sTEC variations between two consecutive epochs:

$$\delta sTEC(t+1,t) = \frac{f_1^{\,2} f_2^{\,2}}{A\left(f_1^{\,2} - f_2^{\,2}\right)} \left[L_{GF}(t+1) - L_{GF}(t)\right] \tag{2}$$

where $A = 40.3 \cdot 10^{16}\ [m][Hz]^2[TECU]^{-1}$ is the standard conversion factor linking TEC [TECU] to ionospheric delay in metric unit [meters]. The discrete derivative of $sTEC$ over time can be simply computed dividing $\delta sTEC$ by the interval between epochs t and $(t+1)$. $sTEC$ is an integrated quantity representing the total number of electrons included in a column with a cross-sectional area of $1\ m^2$, counted along the signal path s between the satellite S and the receiver R. The sTEC observations are modeled by collapsing them to the ionospheric pierce point (IPP) between the satellite-receiver line-of-sight and the single-shell layer located above the height of F2 peak, where the electron density is assumed to be maximum. The IPP position can be computed in real time using standard GNSS broadcast orbit parameters [5], after having chosen the height of the F2 peak.

In this work, single-shell ionospheric layer approximation was applied to explain the physical meaning of the $\delta sTEC$ values provided by VARION and to explicitly show the effect of the IPP motion in the VARION observation equation. This single-shell ionospheric approximation means that the ionospheric sTEC is assigned to an IPP point which renders a 2D picture without vertical dependence of any parameter. In this 2D representation of the ionosphere, the variation $\delta sTEC$ in the interval δt is equivalent to a total derivative over time where the observational point (IPP) moves independently of the motion of the medium (ionospheric plasma). The total derivative encompasses both the variation in time in a certain fixed position (sTEC partial time derivative) and the variation in time due to the sTEC horizontal spatial variation and to the horizontal motion of the IPP relative to the horizontal plasma flow (the relative IPP velocity times the sTEC 2D space gradient on the ionospheric layer); therefore the VARION-MEO (hereafter called VARION-GNSS):

$$\frac{d\, sTEC(t,s)}{dt} = \frac{\partial\, sTEC(t,s)}{\partial t} + \left(\vec{V}_{pla} - \vec{V}_{ipp}\right) \cdot \nabla sTEC(t,s), \tag{3}$$

where $\vec{V}_{pla}$ and $\vec{V}_{ipp}$ are the horizontal plasma and IPP vector velocity field in an Earth-centered Earth-fixed (ECEF) reference frame (WGS84, in our case, since we are using broadcast orbits), respectively, and $\nabla sTEC(t,s)$ is the horizontal spatial gradient of sTEC. It is clear that the convective derivative term accounts for IPP motion and plasma motion $(\vec{V}_{pla} - \vec{V}_{ipp})$. It is important to underline that in a full 3D representation of the ionosphere, $\vec{V}_{pla}$, $\vec{V}_{ipp}$, and $\nabla sTEC(t,s)$ are altitude-dependent terms; in our 2D single-shell ionospheric layer approximation, all these terms are referred to a 300 km height. However, for the purpose of this paper, Eq. (3) already shows that the ionospheric remote sensing based on GNSS observations acquired from MEO satellites depends on the time-dependent position of the IPPs. It is crucial to underline that the $\vec{V}_{ipp}$ magnitude is not constant during the period of observation, but it increases for lower elevation angles [56]. In [53] it was shown that the $\vec{V}_{ipp}$ magnitudes range 40–120 m/s for elevation angles 30–90 degrees, meaning that these IPPs have a velocity of the same order of magnitude of most of the ionospheric perturbations induced by natural hazards (e.g., tsunami-induced TIDs). Also, the background noise, and long period trends of $\delta sTEC$ in

Eq. (3), increases for lower elevation angles, when the length of the signal path inside the ionosphere is longer, leading to larger $\delta sTEC$ values. This explains the current limitation of GNSS ionospheric-based early warning algorithms for low elevation angles. In particular, it is a common practice to apply a cutoff elevation angle for GNSS ionospheric remote sensing studies which is much higher (20 degrees or higher) than the one normally used for GNSS positioning applications (5 degrees or lower).

After identifying and removing cycle slips from $\delta sTEC$ time series, we integrate Eq. (3) over time in order to reconstruct the final $\Delta sTEC$ perturbation term. The VARION approach overcomes the problem of estimating the phase initial ambiguity and the satellite inter-frequency biases (IFBs), which can be assumed constant for a given period [5], thus being ideal for real-time applications.

3.2 VARION-geo

A GEO satellite experiences libration only (i.e., drifting back and forth between two stable points), so that it can be considered motionless relative to an ECEF reference frame, and as a result the IPP's velocity vector $\vec{V}_{ipp}$ is negligible. For this reason, Eq. (3) becomes:

$$\frac{d\,sTEC(t,s)}{dt} = \frac{\partial\,sTEC(t,s)}{\partial t} + \vec{V}_{pla} \cdot \nabla sTEC(t,s), \tag{4}$$

which can be considered the new VARION-GEO observable. Eq. (4) formally reveals the fundamental property of GEO satellites: independence of the estimated $\delta sTEC$ value on the motion of the IPP. Since GEO observations have a constant elevation angle, we can assume a constant level of observational noise throughout the entire period of observation. Furthermore, GEO observations are less prone to trends induced at low elevation angles, when the length of the signal path inside the ionosphere is longer, leading to larger $\delta sTEC$ values. The other important advantage of GEO satellites is the fact that they provide long-term continuous time series over a fixed location.

4. Main results

In this section, we will give an outlook on the main results achieved through the VARION approach. In particular, we will show the main results from [5] for tsunami-generated TID detection (Section 4.1) and from [53] for ionospheric plasma depletion analysis (Section 4.2). For more details on these test cases and on the related data processing performed with VARION, please refer to the cited papers.

4.1 Haida Gwaii tsunami-induced TIDs

4.1.1 Dataset

Using the VARION algorithm, we compute TEC variations induced by the 2012 Haida Gwaii tsunami event at 56 GPS receivers from Plate Boundary Observatory (PBO) in Hawaiian Islands. All the GPS permanent stations are located in Big Island (see **Figure 1**) and acquired observations at 15 and 30 second rate. In order to validate the methodology, results were, hence, compared with the real-time

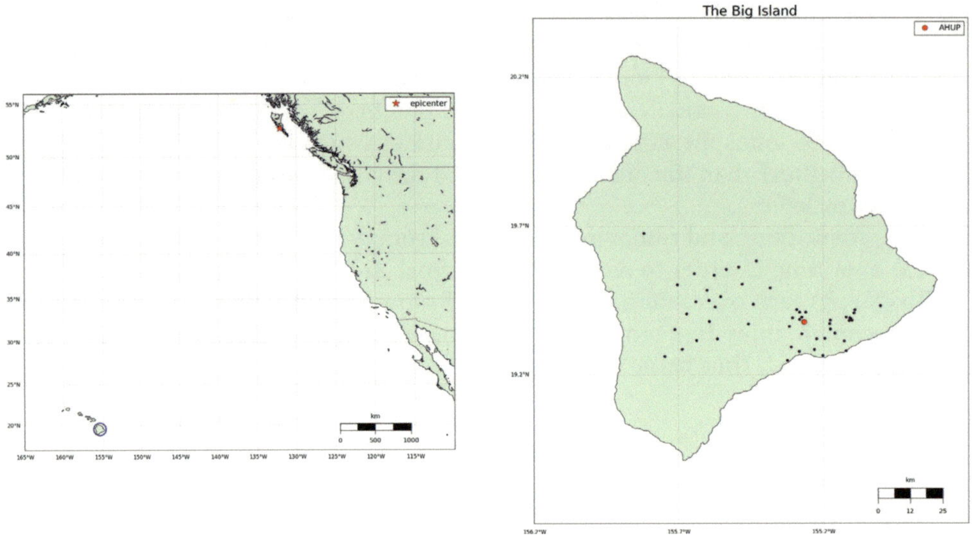

Figure 1.
Map indicating the epicenter of the 10/27/2012 Canadian earthquake (left panel) and zoomed-in image of the Hawaii big island, where the 56 used GPS stations are located. Figure adapted from Savastano et al. [5].

tsunami Method of Splitting Tsunami (MOST) model produced by the NOAA Center for Tsunami Research [57, 58].

4.1.2 Results and discussion

VARION processing outcame a TEC perturbation with amplitudes of up to 0.25 TEC units and traveling ionospheric perturbations (TIDs) moving away from the earthquake epicenter at an approximate speed of 277 m/s. To better study the localized variations of power in the TEC time series, a Paul wavelet analysis was performed [59, 60]. We find perturbation periods consistent with a tsunami typical deep ocean period. In particular, periods in the range of 10–30 min were obtained: these periods are similar to the ones of the tsunami ocean waves, which can range from 5 min up to an hour with the typical deep ocean period of only 10–30 wavelengths around 400 km and the velocity approximately 200 m/s.

Figure 2 shows the sTEC time series wavelet analysis for the seven satellites in view at the station AHUP. The upper panels show the sTEC time series obtained with the VARION software in a real-time scenario. The bottom panels indicate the wavelet spectra. The colors represent the intensity of the power spectrum, and the black contour encloses regions of greater than 95% of confidence for a red noise process. We can identify five satellites (PRNs 4, 7, 8, 10, 20) with peaks consistent in time and period with the tsunami ocean waves. These results clearly show TIDs appearing after the tsunami reached the islands, with an increase of the power spectrum for periods between 10 and 30 min during the TIDs.

Figure 3 shows time sTEC variations for 2 h (08:00–10:00 UT – 28 October 2012) at the IPPs vs. distance from the Haida Gwaii earthquake epicenter, for the same seven satellites under consideration. The TIDs are clearly visible in the interval of significant sTEC variations (from positive to negative values and vice versa). The vertical and horizontal black lines represent the time (when the tsunami arrived at the Hawaii Islands) and the distance (between the epicenter and the Big Island), respectively. In this way, we identify the green rectangle as the alert area, and it is evident that satellite PRN 10, the closest to the earthquake epicenter, detected TIDs before the tsunami arrived at Hawaiian Islands (08:30:08 UT).

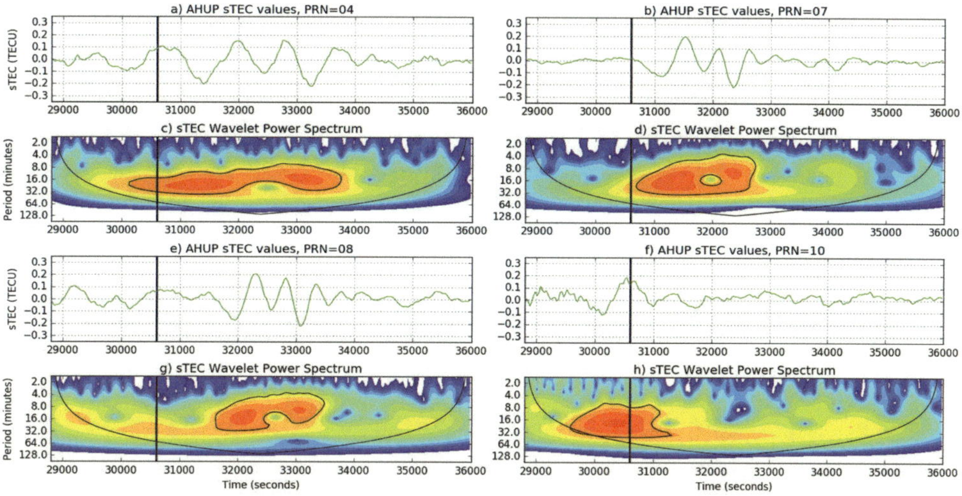

Figure 2.
(a), (b), (e), (f) Four of 260 time series used for the wavelet analysis, station AHUP, satellite PRN 4, 7, 8, 10. (c), (d), (g), (h) the wavelet power spectrum used the Paul wavelet. The vertical axis displays the Fourier period (min), while the horizontal axis is time (s). The black vertical line represents the time when the tsunami reached the Hawaiian islands. The color panels represent the intensity of the power spectrum; the black contour encloses regions of greater than 95% confidence for a red noise process with a lag-1 coefficient of 0.72; the external black line indicates the cone of influence, the limit outside of which edge effects may become significant. Figure adapted from Savastano et al. [5].

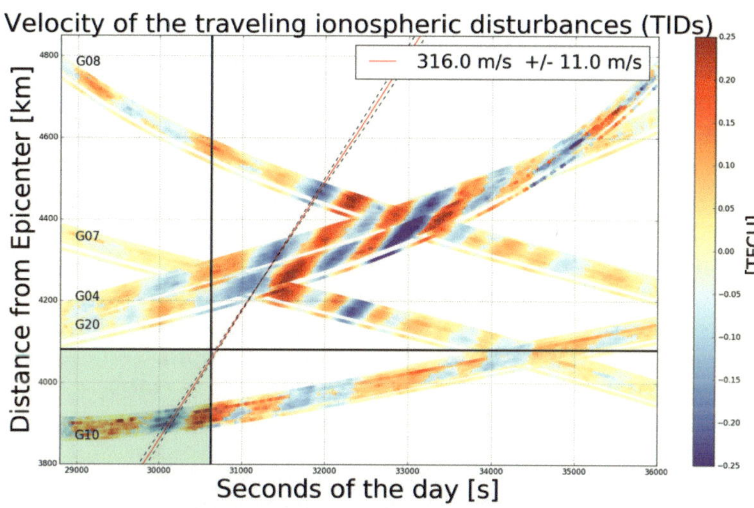

Figure 3.
sTEC variations for 2 h (08:00–10:00 UT – 28 October 2012) at the IPPs vs. distance from the Haida Gwaii earthquake epicenter, for the 7 satellites observed from the 56 Hawaii big islands GPS permanent stations. The TIDs are clearly visible in the interval of significant sTEC variations (from positive to negative values and vice versa). The vertical and horizontal black lines represent the time (when the tsunami arrived at the Hawaii Islands) and the distance (between the epicenter and the big island), respectively; it is evident that PRN 10 detected TIDs before the tsunami arrived at Hawaii Islands (08:30:08 UT). The slope of the straight line fitted, considering a linear least squares regression for corresponding sTEC minima for different satellites, represents the TIDs' mean propagation velocity. Figure adapted from Savastano et al. [5].

In the distance vs. time plots (also called hodochrons), the slope of the straight line, fitted considering corresponding sTEC minima for different satellites, represents the horizontal speed estimate of TIDs. This plot indicates that the linear least squares' estimated speed of the TIDs is about 316 m/s, and it is found to be in good

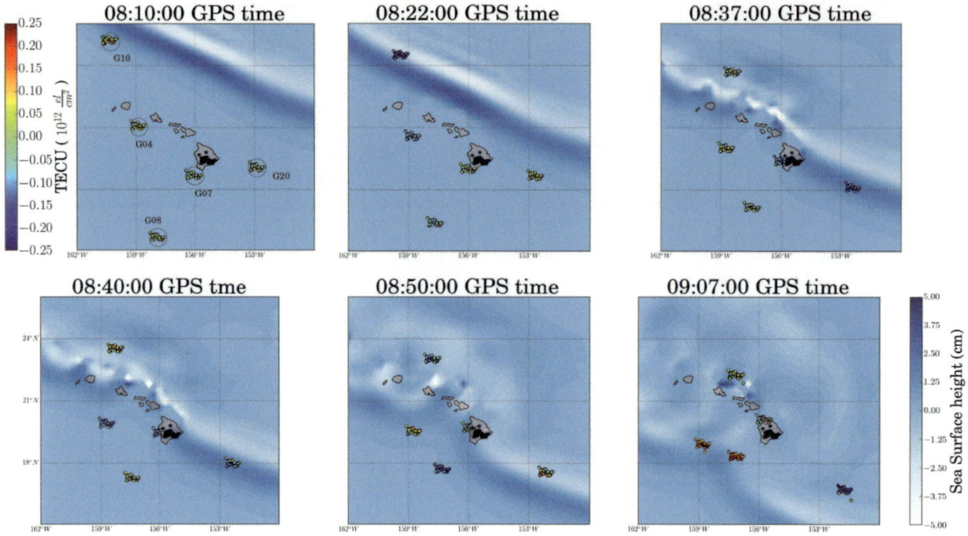

Figure 4.
Space–time sTEC variations at six epochs within the 2-h interval (08:00–10:00 UT – 28 October 2012) at the SIPs for the five satellites showing TIDs over-plotted the tsunami MOST model. TIDs are consistent in time and space with the tsunami waves. Figure adapted from Savastano et al. [5].

agreement with a typical speed of the tsunami gravity waves estimated with ground-based GNSS receivers.

Figure 4 displays a sequence of maps of the region around the Hawaiian Islands showing the variations in sTEC (determinable in real time) at IPP/SPI locations on top of the MOST model sea-surface heights. Note that, just as the MOST model wavefronts are moving past the IPPs, the sTEC variations in the region become pronounced, correlated with the passage of the ocean tsunami itself. In particular, at 08:22:00 GPS time (08:21:44 UT), we are able to see sTEC perturbations from 56 stations looking at satellite PRN 10. The propagation of the MOST modeled tsunami passes the ionospheric pierce points located NW of the Big Island and offers insight with regard to the ionospheric response to the tsunami-driven atmospheric gravity wave. These perturbations are detected before the tsunami reached the islands as seen from the locations of the SIP points. The following frames indicate the tsunami-driven TIDs detected from the other four satellites (PRNs 4, 7, 8, 20) tracking the propagating tsunami (see supplementary video SV1 online).

4.2 Falcon 9 rocket-induced ionospheric plasma depletion

4.2.1 Dataset

To estimate the slant TEC variations associated with the rocket launch, we applied the VARION algorithm to the WAAS-GEO observations collected at 62 Plate Boundary Observatory (PBO) sites located in California (https://www.unavco.org/instrumentation/networks/status/pbohttps://www.unavco.org/instrumentation/networks/status/pbo). In this study, we used satellite S35 (PRN 135) located at 133 degree West and satellite S38 (PRN 138) located at 107.3 degree West. **Figure 5** (left panel) shows the IPP location for satellites S35 (blue dots) and S38 (yellow dots) and the location of the ionosonde site PA836 (red dot). We use the standard single-shell ionospheric layer approximation at the height of 300 km to calculate the IPP locations [61]. On the right, two maps representing the Earth as seen from these two GEO satellites are shown. The raw GEO observations are available in RINEX format with a sampling rate of 15 seconds.

The ionosonde observations from site PA836 (located less than 5 kilometers from the Vandenberg Air force Base) are used here for an independent comparison with the VARION-GEO solutions. The electron density profiles derived from the sweeping ionosonde observations extend from the lower E region to the F region peak with 15 min of cadence.

4.2.2 Results and discussion

Figure 6(a) shows the closest VARION-GEO Δ*sTEC* time series to the ionosonde site and (b) shows the ionosonde peak electron density (NmF2)

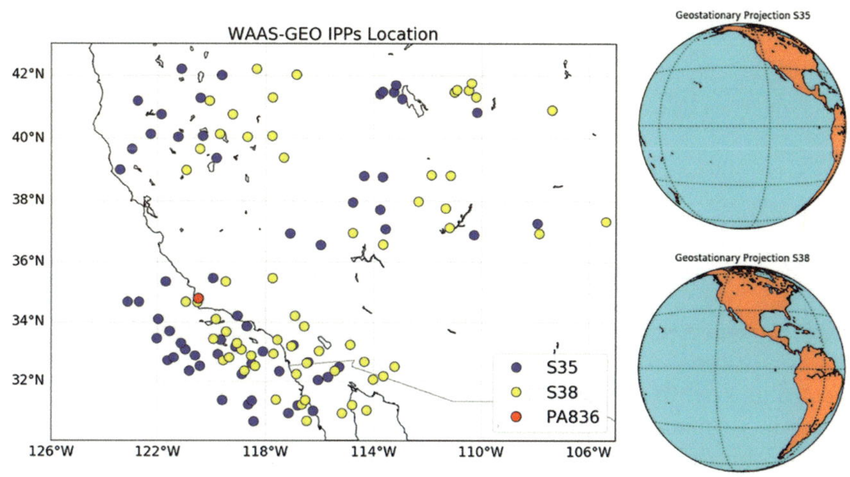

Figure 5.
Map showing the IPP location for satellites S35 (blue dots) and S38 (yellow dots) seen from the 62 GNSS stations. The IPPs for GEO satellites can be considered to be fixed over time. The red dot represents the location of the ionosonde site PA836. On the right, we display two maps representing the earth as seen from WAAS-GEO satellites S35 and S38. Figure adapted from Savastano et al. [53].

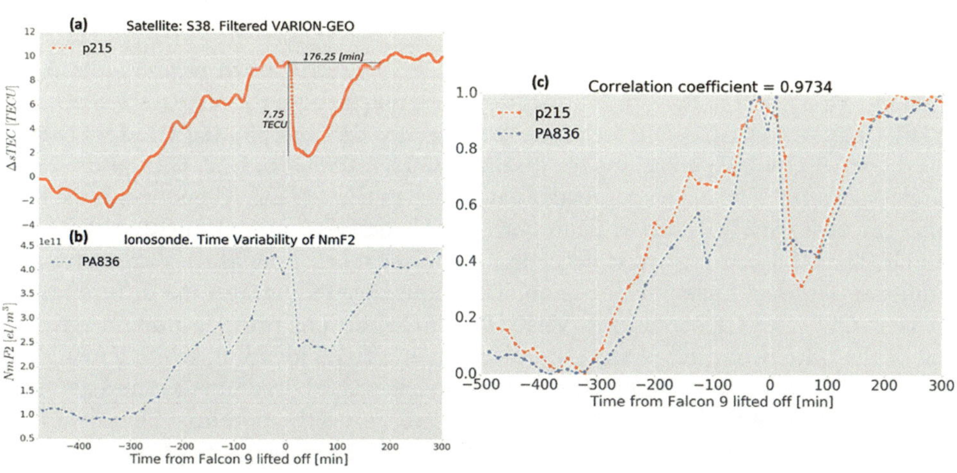

Figure 6.
(a) Shows the VARION-GEO ΔsTEC solutions obtained from station p215, satellite S38. (b) Shows the NmF2 time variability obtained from ionosonde PA836. (c) Shows the down-sampled and normalized ΔsTEC solutions (red curve) and the normalized NmF2 time series (blue curve) plotted using a common scale [0, 1]. This figure shows a high correlation between the VARION-GEO ΔsTEC solutions and ionosonde data. The correlation coefficient between the two curves is 0.97. Figure adapted from Savastano et al. [53].

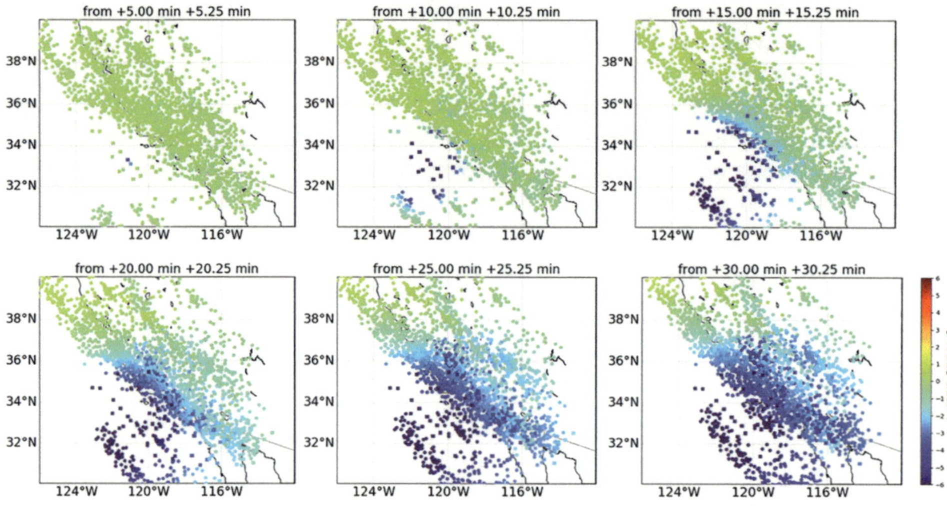

Figure 7.
Space–time ΔsTEC variations for 30 min after the launch (one frame every 5 min) at the SIPs (same positions of the corresponding IPPs on the map) for the 2 GEO satellites (square symbols) and 6 GPS satellites (denoted by circles) seen from the 62 GNSS permanent stations. The ionospheric hole is detected from both GEO satellites 5 min after the rocket launch. The coordinates are expressed in geodetic latitude (in degrees north) and longitude (in degrees west). Figure adapted from Savastano et al. [53].

extracted for each electron density profile measured by the ionosonde and plotted as a function of time. An electron density depletion in the F2 layer is clearly visible from both data sets. In order to quantify the agreement between the two curves, we applied a min-max normalization to the two curves to bring all values into the range [0, 1]. This procedure allows us to study the correlation between the two curves: **Figure 6(c)** displays the normalized *ΔsTEC* (red) and NmF2 (blue) curves. We then down-sampled the normalized VARION-GEO solutions in order to have the same sampling rate as the ionosonde data (15 min). Finally, we computed the correlation coefficient between the two curves, and we found a value of 0.97. Despite the fact the ionosonde electron density profiles extend up to the F2 peak, and that two measurements are not exactly co-located, the agreement between the two datasets is very good.

Figure 7 displays a sequence of six maps (every 5 min) in the region around Vandenberg Air Force Base in California. These maps show the VARION-GEO *ΔsTEC* solutions at GEO-IPP locations for satellites S35 and S38 (squared markers) and the VARION-GPS solutions for satellites G02, G05, G06, G12, G25, and G29 (circle markers). The colors represent variations in the *ΔsTEC*. The ionospheric hole (blue color) is clearly detected from both GEO satellites 5 min after the rocket launch. The GPS satellites start detecting the ionospheric hole as they are moving inside the depleted ionospheric region. This figure well illustrates the difference between GEO and GPS solutions. VARION-GEO solutions provide a direct estimation of the time evolution of the ionosphere over a fixed location, while VARION-GPS solutions are also affected by the ionospheric spatial gradients as they move along the IPP trajectory (Section 3.2). The figure shows the potential benefits of GEO satellites as a complementary technique for well-established GPS satellites.

5. Conclusions

It is widely known that ionospheric anomalies can be a threat to GNSS and satellite telecommunications; therefore real-time monitoring of the ionosphere represents an important outreach. This chapter finds its reasons in this background, but in the meantime it extends its fields of application to natural hazard early warning systems. It represents an overview about the possible real-time VARION applications for the monitoring of ionospheric irregularities and TEC perturbations.

The VARION is based on single time difference of geometry-free combination of carrier-phase observations that makes it suitable for real-time application. The VARION algorithm was applied both to standard GNSS MEO satellites and to GNSS GEO satellites. It is important to underline that these analyses were carried out in real-time scenario: only data available in real time were used.

In detail, the 2012 Haida Gwaii tsunami event represents a fundamental study case as it showed for the first time that real-time detection of tsunami-induced TEC perturbations is possible and that these TIDs become clear before the tsunami waves hit the Hawaii Big Island [5]. This paper demonstrated that real-time GNSS tracking of TEC perturbations can provide information on tsunami propagation that is consistent with that generated by NOAA's current real-time forecast system [62]. The ability of VARION to detect the TIDs before the tsunami arrival represents a valid contribution for the enhancement of tsunami early warning system.

In [53], it was demonstrated that the extension of the VARION algorithm to GEO satellites enabled a better description of the ionospheric plasma depletion induced by a Falcon 9 rocket. These results are relevant for different GNSS applications, since an ionospheric plasma depletion can potentially lead to a range error of several meters. Lastly, the VARION was implemented in the JPL's Global Differential GPS System (GDGPS) real-time interface that may be accessed at (https://iono2la.gdgps.net/), allowing real-time monitoring of the status of the ionosphere.

Therefore, the VARION extreme versatility makes it suitable for real-time ionospheric monitoring and anomaly detection applications.

Acknowledgements

The authors thank Prof. Mattia Crespi for his great support throughout of the drawing up of this chapter.

Conflict of interest

The authors declare no conflict of interest.

Abbreviations

IPP	ionospheric pierce point
SIP	sub-ionospheric pierce point
VARION	Variometric Approach for Real-Time Ionosphere Observation
TIDs	traveling ionospheric disturbances
CIDs	coseismic ionospheric disturbances
MEO	medium Earth orbit
GEO	geostationary orbit
AGWs	atmospheric gravity waves
SAWs	shock acoustic waves
TEC	total electron content
PBO	Plate Boundary Observatory network
WAAS	Wide Area Augmentation System
MOST	Method of Splitting Tsunami

Author details

Giorgio Savastano[1*†] and Michela Ravanelli[2*†]

1 Spire Global, Inc., Luxembourg

2 Geodesy and Geomatics Division – DICEA, Sapienza University of Rome, Rome, Italy

*Address all correspondence to: giorgio.savastano@spire.com
and michela.ravanelli@uniroma1.it

† These authors contributed equally.

IntechOpen

© 2019 The Author(s). Licensee IntechOpen. This chapter is distributed under the terms of the Creative Commons Attribution License (http://creativecommons.org/licenses/by/3.0), which permits unrestricted use, distribution, and reproduction in any medium, provided the original work is properly cited.

References

[1] Afraimovich EL, Astafyeva EI, Demyanov VV, Edemskiy IK, Gavrilyuk NS, Ishin AB, et al. A review of GPS/GLONASS studies of the ionospheric response to natural and anthropogenic processes and phenomena. Journal of Space Weather and Space Climate. 2013;**3**:A27

[2] Su K, Jin S, Hoque MM. Evaluation of ionospheric delay effects on multi-gnss positioning performance. Remote Sensing. 2019;**11**(2):171

[3] Larson KM. Unanticipated uses of the global positioning system. Annual Review of Earth and Planetary Sciences. 2019;**47**(1):19-40

[4] Mannucci AJ et al. A global mapping technique for GPS derived ionospheric total electron content measurements. Radio Science. 1998;**33**: 565-582

[5] Savastano G, Komjathy A, Verkhoglyadova O, Mazzoni A, Crespi M, Wei Y, et al. Real-time detection of tsunami ionospheric disturbances with a stand-alone gnss receiver: A preliminary feasibility demonstration. Scientific Reports. 2017;7. Article number: 46607

[6] Liu Z, Yang Z, Chen W. A study on ionospheric irregularities and associated scintillations using multi-constellation gnss observations. In: Proc. of ION PNT, Institute of Navigation, Honolulu, Hawaii, USA. 2017

[7] Kintner PM, Ledvina BM, De Paula ER. GPS and ionospheric scintillations. Space Weather. 2007;**5** (9). Article number: S09003

[8] Masashi Hayakawa and Oleg A Molchanov. Seismo Electromagnetics: Lithosphere-Atmosphere-Ionosphere Coupling. 2002

[9] Occhipinti G, Kherani EA, Lognonné P. Geomagnetic dependence of ionospheric disturbances induced by tsunamigenic internal gravity waves. Geophysical Journal International. 2008; **73**:753765

[10] Afraimovich EL, Boitman ON, Zhovty EI, Kalikhman AD, Pirog TG. Dynamics and anisotropy of traveling ionospheric disturbances as deduced from transionospheric sounding data. Radio Science. 1999;**34**(2):477-487

[11] Afraimovich E. The spatio-temporal characteristics of the wave structure excited by the solar terminator as deduced from TEC measurements at the global GPS network. In: EGU General Assembly Conference Abstracts. Vol. 11. 2009. p. 62

[12] Rolland LM, Lognonné P, Munekane H. Detection and modeling of Rayleigh wave induced patterns in the ionosphere. Journal of Geophysical Research: Space Physics. 2011;**116**(A5). Article number: A05320

[13] Komjathy A, Galvan DA, Stephens P, Butala MD, Akopian V, Wilson B, et al. Detecting ionospheric tec perturbations caused by natural hazards using a global network of gps receivers: The tohoku case study. Earth, Planets and Space. 2013;**64**(12):24

[14] Rolland LM, Lognonné P, Astafyeva E, Kherani EA, Kobayashi N, Mann M, et al. The resonant response of the ionosphere imaged after the 2011 off the Pacific coast of Tohoku earthquake. Earth, Planets and Space. 2011;**63**(7):62

[15] Dautermann T, Calais E, Lognonné P, Mattioli GS. Lithosphere-atmosphere-ionosphere coupling after the 2003 explosive eruption of the Soufriere hills volcano, Montserrat.

Geophysical Journal International. 2009; **179**(3):1537-1546

[16] Kherani EA, Lognonné P, Kamath N, Crespon F, Garcia R. Response of the ionosphere to the seismic triggered acoustic waves: Electron density and electromagnetic fluctuations. Geophysical Journal International. 2009;**176**:1-13

[17] Galvan DA et al. Ionospheric signatures of Tohoku-Oki tsunami of march 11, 2011: Model comparisons near the epicenter. Radio Science. 2012;**47**: RS4003

[18] Astafyeva E, Heki K, Kiryushkin V, Afraimovich E, Shalimov S. Two-mode long-distance propagation of coseismic ionosphere disturbances. Journal of Geophysical Research: Space Physics. 2009;**114**(A10). Article number: A10307

[19] Galvan DA, Komjathy A, Hickey MP, Mannucci AJ. The 2009 Samoa and 2010 Chile tsunamis as observed in the ionosphere using GPS total electron content. Journal of Geophysical Research. 2011;**116**:A06318

[20] Afraimovich EL, Perevalova NP, Plotnikov AV, Uralov AM. The shock-acoustic waves generated by the earthquakes. Annales de Geophysique. 2001;**19**:395-409

[21] Ducic V, Artru J, Lognonné P. Ionospheric remote sensing of the Denali earthquake Rayleigh surface waves. Geophysical Research Letters. 2003;**30**(18)

[22] Artru J, Farges T, Lognonné P. Acoustic waves generated from seismic surface waves: Propagation properties determined from Doppler sounding observations and normal-mode modelling. Geophysical Journal International. 2004;**158**(3):1067-1077

[23] Tahira M. Acoustic resonance of the atmospheric at 3.7 hz. Journal of the Atmospheric Sciences. 1995;**52**(15): 2670-2674

[24] Queney P. The problem of airflow over mountains. A summary of theoretical studies. Bulletin of the American Meteorological Society. 1948; **29**:16-26

[25] Rakoto V, Lognonné P, Rolland L, Coisson P. Tsunami wave height estimation from GPS-derived Ionospheric data. Journal of Geophysical Research: Space Physics. 2018;**123**(5): 4329-4348

[26] Daniels FB. Acoustic energy generated by ocean waves. Journal of the Acoustical Society of America. 1952; **24**:83

[27] Hines CO. Internal atmospheric gravity waves at ionospheric heights. Canadian Journal of Physics. 1960;**38**: 14411481

[28] Hines CO. Gravity waves in the atmosphere. Nature. 1972;**239**:7378

[29] Peltier WR, Hines CO. On the possible detection of tsunamis by a monitoring of the ionosphere. Journal of Geophysical Research. 1976;**81**(12): 1995-2000

[30] Rolland LM, Occhipinti G, Lognonné P, Loevenbruck A. Ionospheric gravity waves detected offshore Hawaii after tsunamis. Geophysical Research Letters. 2010;**37**: L17101

[31] Occhipinti G, Lognonné P, Kherani EA, Hbert H. Three-dimensional waveform modeling of ionospheric signature induced by the 2004 Sumatra tsunami. Geophysical Research Letters. 2006; **33**(20). Article number: L20104

[32] Lee MC et al. Did tsunami-launched gravity waves trigger ionospheric

turbulence over Arecibo. Journal of Geophysical Research. 2008;**113**:A01302

[33] Coisson P, Lognonné P, Walwer D, Rolland LM. First tsunami gravity wave detection in ionospheric radio occultation data. Earth and Space Science. 2015;**2**:125-113

[34] Garcia RF, Doornbos E, Bruinsma S, Hebert H. Atmospheric gravity waves due to the Tohoku-Oki tsunami observed in the thermosphere by GOCE. Journal of Geophysical Research-Atmospheres. 2014;**119**(8):4498-4506

[35] Kelley MC. In situ ionospheric observations of severe weather-related gravity waves and associated small-scale plasma structure. Journal of Geophysical Research: Space Physics. 1997;**102**(A1): 329-335

[36] Nicolls MJ, Kelley MC, Coster AJ, Gonzalez SA, Makela JJ. Imaging the structure of a large scale tid using isr and tec data. Geophysical Research Letters. 2004;**31**:L09812

[37] Afraimovich EL, Astafyeva EI, Demyanov VV, Gamayunov IF. Mid-latitude amplitude scintillation of gps signals and gps performance slips. Advances in Space Research. 2009; **43**(6):964-972

[38] Calais E, Minster JB. GPS detection of ionospheric perturbations following the January 17, 1994, Northridge earthquake. Geophysical Research Letters. 1995;**22**(9). 1045-1048

[39] Artru J, Lognonné P, Blanc E. Normal modes modelling of postseismic ionospheric oscillations. Geophysical Research Letters. 2001;**28**(4):697-700

[40] Kelley MC, Livingston R, McCready M. Large amplitude thermospheric oscillations induced by an earthquake. Geophysical Research Letters. 1985;**12**:577-580

[41] Artru J, Ducic V, Kanamori H, Lognonné P, Murakami M. Ionospheric detection of gravity waves induced by tsunamis. Geophysical Journal International. 2005;**20**:840-848

[42] Crowley G, Rodrigues FS. Characteristics of traveling ionospheric disturbances observed by the tiddbit sounder. Radio Science. 2012;**47**(4):1-12

[43] Ogawa T, Igarashi K, Aikyo K, Maeno H. NNSS satellite observations of medium-scale traveling Ionospheric disturbances at southern high-latitudes. Journal of Geomagnetism and Geoelectricity. 1987;**39**(12):709-721

[44] Booker HG. A local reduction of f-region ionization due to missile transit. Journal of Geophysical Research. 1961; **66**(4):1073-1079

[45] Mendillo M, Hawkins GS, Klobuchar JA. A sudden vanishing of the ionospheric F region due to the launch of Skylab. Journal of Geophysical Research. 1975;**80**:2217-2225

[46] Bernhardt PA, Huba JD, Kudeki E, Woodman RF, Condori L, Villanueva F. Lifetime of a depression in the plasma density over Jicamarca produced by space shuttle exhaust in the ionosphere. Radio Science. 2001;**36**: 1209-1220

[47] Bernhardt PA et al. Ground and space-based measurement of rocket engine burns in the ionosphere. IEEE Transactions on Plasma Science. 2012; **40**:1267-1286

[48] Mendillo M, Smith S, Coster A, Erickson P, Baumgardner J, Martinis C. Man-made space weather. Space Weather. 2008;**6**:S09001

[49] Ozeki M, Heki K. Ionospheric holes made by ballistic missiles from North Korea detected with a Japanese dense GPS array. Journal of Geophysical

Research: Space Physics. 2010;**115**(A9). Article number: A09314

[50] Komjathy A, Sparks L, Wilson BD, Mannucci AJ. Automated daily processing of more than 1000 ground-based GPS receivers for studying intense ionospheric storms. Radio Science. 2005;**40**:RS6006

[51] Sardon E, Rius A, Zarraoa N. Estimation of the transmitter and receiver differential biases and the ionospheric total electron content from global positioning system observations. Radio Science. 1994;**29**:577-586

[52] Hajj GA, Lee LC, Pi X, Romans LJ, Schreiner WS, Straus PR, et al. COSMIC GPS ionospheric sensing and space weather. Terrestrial, Atmospheric and Oceanic Sciences. 2000;**11**:235-272

[53] Savastano G, Komjathy A, Shume E, Vergados P, Ravanelli M, Verkhoglyadova O, et al. Advantages of geostationary satellites for ionospheric anomaly studies: Ionospheric plasma depletion following a rocket launch. Remote Sensing. 2019;**11**(14):1734

[54] Bishop G, Walsh D, Daly P, Mazzella A, Holland E. Analysis of the temporal stability of GPS and GLONASS group delay correction terms seen in various sets of ionospheric delay data. In: Proceedings of the 7th International Technical Meeting of the Satellite Division of the Institute of Navigation (ION GPS 1994). 1994. pp. 1653-1661

[55] Demyanov VV, Yasyukevich YV, Jin S, et al. The second-order derivative of gps carrier phase as a promising means for ionospheric scintillation research. Pure and Applied Geophysics. 2019;**176**(10):4555-4573

[56] Savastano G. New applications and challenges of GNSS variometric approach [Ph.D. dissertation]. Rome, Italy: Dept. DICEA, Univ. La Sapienza; 2018

[57] Wei Y et al. Real-time experimental forecast of the Peruvian tsunami of august 2007 for U.S. coastlines. Geophysical Research Letters. 2008;**35**: L04609

[58] Wei Y, Chamberlin C, Titov V, Tang L, Bernard EN. Modeling of the 2011 Japan tsunami - lessons for near-field forecast. Pure and Applied Geophysics. 2013;**170**(6–8):1309-1331

[59] Torrence C, Compo GP. A practical guide to wavelet analysis. Bulletin of the American Meteorological Society. 1998; **79**:6178

[60] Misiti M, Misiti Y, Oppenheim G, Poggi J-M. Wavelet Toolbox. Vol. 15. Natick, MA: The MathWorks Inc.; 1996. p. 21

[61] Klobuchar JA. Ionospheric time-delay algorithm for single-frequency GPS users. IEEE Transactions on Aerospace and Electronic Systems. 1987; **AES-23**(3):325-331

[62] Murray JR, Bartlow N, Bock Y, Brooks BA, Foster J, Freymueller J, et al. Regional global navigation satellite system networks for crustal deformation monitoring. Seismological Research Letters. 2019

Section 2

Impacts on GNSS: Modeling and Mitigation Techniques

Chapter 4

GPS Signal Multipath Error Mitigation Technique

Bharati Bidikar, Babji Prasad Chapa,
Mogadala Vinod Kumar and Gottapu Sasibhushana Rao

Abstract

The performance of GPS receiver depends on the accuracy of the range measurements. The predominant errors in range measurements are due to propagation path delays, making the measured range longer than it would be, if the signal has not reflected or refracted while propagating. In this chapter, an algorithm is proposed to mitigate the multipath error on the pseudorange measured from L1 carrier frequency. The error is estimated considering the linear combination of the GPS measurements and carrier frequencies of L band, viz. L1 and L2. This algorithm exploits the random nature of the multipath error and it avoids complex calculations involving sensitive parameter like reflection coefficient of the nearby reflectors. The multipath error is mitigated for standalone GPS receiver located in Indian subcontinent. Implementation of the algorithm shows pseudorange error due to multipath varied from 7 to 52 m, where the signals of low elevation satellites are most affected. GPS receiver position is calculated by considering multipath error corrected pseudoranges of all the visible satellites. This resulted in maximum error reduction of 30 m in receiver position estimates. This mitigation technique will be useful in selecting the site for GPS receiving antenna, where reflection coefficients are difficult to measure.

Keywords: GPS, L1 and L2 frequencies, elevation angle, pseudorange, carrier phase range, multipath error, ionospheric delay

1. Introduction

GPS finds its applications in most of the day to day activities of human life, viz. precise farming, surveying, missile guidance, military and civil aviation [1]. However, the accuracy, availability, reliability, and integrity of GPS navigation solution are impaired by various errors which are originating at the satellites, like orbital errors, satellite clock errors, etc. [2]; whereas, the receiver clock errors, multipath errors, receiver noise, and antenna phase center variations are the errors originating at the receiver [3]. Also, the propagation medium contributes to the delays in the GPS signal, as it passes through the ionosphere and troposphere [4]. In addition to these errors, the accuracy of the navigation solution is also affected by GPS satellites location as viewed by the receiver. Hence, error estimation and correction is a primary concern in precise navigation applications. In this chapter,

IntechOpen

the error originating at the receiver which is due to GPS signal multiple paths is addressed.

The signal transmitted by the satellite, taking multiple paths, affects both the pseudorange and carrier phase measurements [5]. But the effect of multipath on pseudorange is much higher than the carrier phase [6, 7]. The pseudorange multipath error in an urban environment is characterized by considering signal to noise ratio and elevation angle for DGPS [8]. The multipath effect on carrier phase measurements was also detected [9]. The GPS receiver cannot distinguish the direct signal from the several multipath signals. This problem in tracking loop was addressed by several authors [10, 11]. For the same GPS receiver, multipath error differs depending on the reflecting surfaces, viz. multipath effect due to large water bodies like sea surface. Multipath due to water bodies and its impact on the precision of GNSS positioning in marine application was also studied [12] and similar studied were done for static and kinematic receivers [13]. There are several studies mitigating the multipath error by antenna based mitigation methods. A choke ring antenna with ground plane to absorb multipath signals was proved to mitigate the error to large extent [14, 15]. Mitigation of the error and the performance of the receivers are analyzed for dual frequency receiver as well [16]. Some techniques also rely on the analysis of the signal-to-noise ratio values of GPS signals [17]. Apart from these methods the filter-based techniques are also implemented to extract or eliminate multipath effects, such as wavelet filters [18–20], Vondrak filter [21] and adaptive filter [22]. In precise positioning applications, multipath is a major error source and impact needs to be calculated especially in urban canyon while setting up GPS receiver antenna [23]. The multipath error originating at the receiver is very sensitive to geometry (like size and surface texture) and the reflection coefficients of the nearby reflectors [24]. These parameters limit the efficiency of the conventional multipath modeling methods. But in this chapter, an algorithm is proposed to calculate the multipath error affecting GPS L1 pseudorange range measurement. The algorithm utilizes the relationship between the code range measurements, carrier phase measurements and carrier frequencies (L1 = 1575.42 MHz and L2 = 1227.60 MHz) [25]. In this chapter, the multipath error affecting the pseudorange measurements of Satellite Vehicle Pseudorandom Noise (SVPRN) 07, 23, 28 and 31 are estimated using the proposed algorithm. The ephemerides data of these satellites for the entire day was collected on March 11, 2019, from a standalone GPS receiver, located in the Indian subcontinent. The proposed algorithm and the impact analysis done in this chapter will also be a valuable aid for setting up the GPS receiver antenna for air traffic control and navigation. Section 2 briefly explains the error budget and explains multipath error in detail. It also gives brief overview of the existing multipath error mitigation techniques. Section 3 gives the proposed multipath error estimation algorithm. Finally, the results and conclusions are given in Sections 4 and 5 respectively.

2. Multipath error in GPS

GPS measurements are biased by many errors. These errors are specific to each satellite signal and translate into a receiver position error (the receiver position being calculated from the estimated travel time of the signal from each satellite to the receiver). The errors are divided into three major groups as,

- Errors originating at the satellite like satellite clock error, ephemeris error, and error due to orbital eccentricity

- Errors originating in the propagation medium like ionospheric delay and neutral atmospheric delay
- Errors originating at the receiver like multipath error, receiver clock error, and instrument biases.

The major error sources their impact on PGS range measurements are given in **Table 1** [26].

A GPS signal may take several paths to a receiver's antenna and the signal can be reflected from buildings or ground and interfere with the direct signal creating a range error of several meters or more. The error impact on pseudorange measurement is much larger than that on the carrier phase. The inaccuracy in pseudorange directly affects the receiver position estimation. But for carrier phase measurements, the inaccuracy due to multipath will lead to a wide ambiguity search space and hence takes a longer time to resolve the ambiguity. This will result in incorrect determination of initial ambiguity which further leads to positioning errors. The signal delay due to multipath is very sensitive to the reflection coefficients of the nearby reflectors. These parameters limit the efficiency of the multipath modeling techniques. The impact of the multipath error on GPS satellite signal is given in the following sub section.

2.1 Multipath effect on GPS measurements

GPS receiver determines the pseudorange by code tracking and carrier tracking method. Code tracking method estimates the propagation time and carrier phase tracking method estimates phase delay between the received carrier and the locally generated signal. To measure the propagation time, i.e., the code range measurement, the locally generated code is shifted in time and correlated with the received signal. The correlation parameter is used by a discriminator to adjust the locally generated code with the received code and obtain the time delay. This time delay when scaled by the speed of light gives the range between the satellite and the receiver. A common code discriminator function used in GNSS receivers is the early correlator (E) minus Late correlator (L) values (E-L). The early correlator value is

Error source	Nominal values (m)	Remarks
Orbit	1–5	Error in broadcast ephemeris due to residual errors in curve fitting
Clock	3–5	Due to satellite clock drift
Ionosphere	2–50	Depends upon satellite elevation angle and solar activity
Troposphere	2–30	Depends upon the water vapor content in the lower part of atmosphere
Code multipath	15–150	Maximum 150 m using one chip correlator spacing and 15 m using 0.1 chip correlator spacing
Code noise	0.1–3	For C/A code. Depends upon receiver technology and dynamic stress
Carrier multipath	0.001–0.03	Maximum 4.75 cm for L1 carrier and 6.11 cm for L2 carrier
Carrier noise	0.0002–0.002	For L1 carrier. Depends upon receiver technology and dynamic stress

Table 1.
GPS error sources for SPS receivers.

the correlation between the incoming code and an early version of the locally generated code. The Late correlator value is the correlation value between the incoming code and a late version of the locally generated code. In the presence of multipath, the time delay is estimated by correlating the composite signal with locally generate code(s), which result in code measurement errors. The impact of multipath on code phase measurements can be up to half a chip equivalent to a range error of about 150 m for the GPS C/A code.

In a GPS receiver, the carrier phase is measured by accumulating the phase of the numerically controlled oscillator (NCO) output. In an environment, where there are no reflected signals, the incoming signal carrier is the same as the direct signal carrier. The NCO generated local carrier locks onto the direct carrier very accurately, and, as a result, the true phase difference between the incoming signal carrier and the locally generated carrier is nearly zero, (actually zero mean), at steady state. The resulting phase measurements are very accurate. In the presence of multipath, however, the composite signal phase shifts from the direct signal phase, and the NCO-generated local carrier locks onto the composite carrier phase, resulting in an error in the phase measurement. The processing of the received signal in GPS receiver is shown below.

The change in carrier phase due to multipath effect can be determined in the PLL section of the GPS receiver as shown in **Figure 1**. Following steps are carried out to extract the error in phase due to multipath,

Step 1: The direct signal received at the receiver is $Acos\varphi$. Here the "A" is amplitude and "φ" is phase angle.

Step 2: The reflected signal is modeled as $\alpha(Acos\varphi + \Delta\varphi)$, where "$\alpha$" is attenuation constant and "$\Delta\varphi$" is change in carrier phase of the maximum reflected signal.

Step 3: The composite signal received as the GPS receiver is

$$A\cos\varphi + \alpha(A\cos\varphi + \Delta\varphi) \tag{1}$$

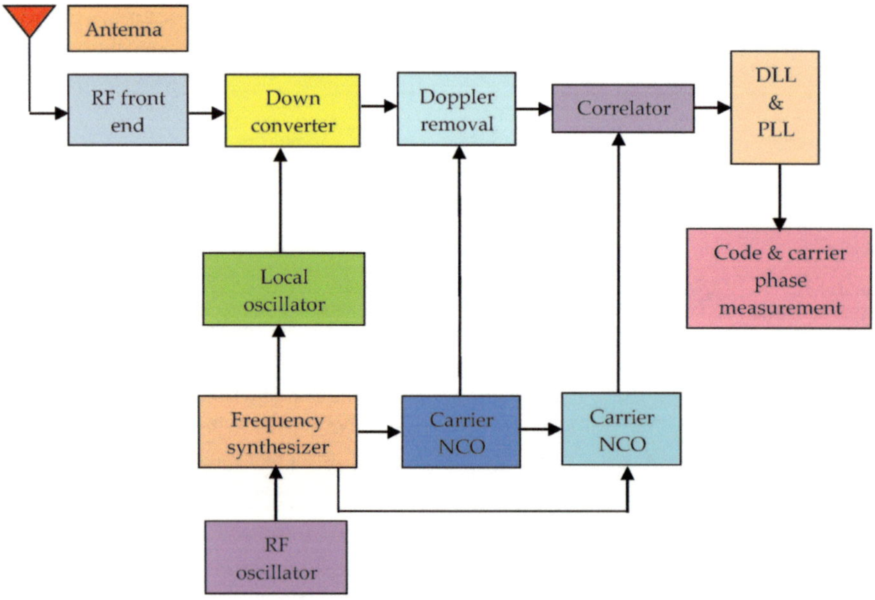

Figure 1.
GPS receiver block diagram showing extraction of code and carrier phase measurements from received composite signal.

Step 4: The error in phase measurement due to multipath is modeled as,

$$\delta\varphi = \tan^{-1}\left(\frac{\sin\,\Delta\varphi}{\alpha^{-1} + \cos\,\Delta\varphi}\right) \tag{2}$$

The above explained method involves tedious trigonometric relationship and is difficult to determine the error due to multipath. To mitigate the multipath error, highly sensitive GPS receivers utilize multiple narrow-spaced correlators. But most of the multipath mitigation techniques related to GPS receiver hardware are not cost effective and need complex hardware to implement; whereas the data processing methods to mitigate the multipath error are more effective. These methods involve correction of GPS code and carrier phase measurements. Multipath elimination delay lock loop are used to mitigate multipath at the receiver signal processing level [27]. Most modern GPS receivers now employ similar algorithms. However, multipath cannot be completely removed and the residuals may still be too large to ignore when high accuracy positioning results are required. The antenna based mitigation techniques involve the use of antenna with a high sensitivity to right-hand circular polarized (RHCP) signals, choke-ring-ground-plane antenna and antenna arrays. The reflected signals typically contain a large LHCP component. Multipath susceptibility of an antenna can be quantified with respect to the antenna's gain pattern characteristics by the multipath ratio (MRP) [28]. Most of the multipath error modeling or mitigation methods are complex to implement; hence in the following section, a multipath mitigation technique on pseudorange on L1 carrier is given, which makes the use of the linear combination of GPS measurements.

3. Pseudorange multipath error mitigation

Among the many errors affecting the GPS measurements [29], the predominant errors like ionospheric delay, multipath error and integer ambiguity are considered in the method. The range measurements on L1 and L2 carrier frequency are given by Eq. (3), (4), and (5).

$$P_{L1} = \rho + I_{L1} + MP_{L1} \tag{3}$$

where P_{L1} = pseudorange on L1 frequency [m]; ρ = geometric range [m]; I_{L1} = ionospheric delay on L1 frequency [m]; MP_{L1} = multipath error on P_{L1} [m].

$$\varphi_{L1} = \rho - I_{L1} + \lambda_{L1}N_{L1} + m\varphi_{L1} \tag{4}$$

$$\varphi_{L2} = \rho - I_{L2} + \lambda_{L2}N_{L2} + m\varphi_{L2} \tag{5}$$

where φ_{L1} = Carrier phase measurement on L1 frequency [m]; φ_{L2}= Carrier phase measurement on L2 frequency [m]; N_{L1}, N_{L2}= Integer ambiguity on L1 and L2 frequencies respectively; λ_{L1}=Wavelength of L1 carrier frequency [m]; λ_{L2}=Wavelength of L2 carrier frequency [m]; $m\varphi_{L1}$ = Multipath error on φ_{L1}[m]; $m\varphi_{L2}$= multipath error on φ_{L2} [m].

The multipath error in carrier phase measurements ($m\varphi_{L1}$ and $m\varphi_{L2}$) are assumed to be negligible compared to the error in pseudorange measurement. The expression for MP_{L1} can be obtained by forming the appropriate linear combination of code range and carrier phase measurements (subtract Eq. (4) from Eq. (3)).

$$\begin{aligned} P_{L1} - \varphi_{L1} &= 2I_{L1} - \lambda_{L1}N_{L1} + MP_{L1} \\ P_{L1} - \varphi_{L1} - 2I_{L1} &= MP_{L1} - \lambda_{L1}N_{L1} \end{aligned} \tag{6}$$

To make the above equation free from ionospheric delay, the Eq. (5) is subtracted from Eq. (4) and rearranged as below,

$$\varphi_{L1} - \varphi_{L2} = I_{L2} - I_{L1} + \lambda_{L1} N_{L1} - \lambda_{L2} N_{L2} \tag{7}$$

Since the ionosphere is dispersive medium, the delay is frequency dependent. Hence the delays (I_{L1} and I_{L2}) are related to the respective carrier frequencies (f_{L1} and f_{L2}) as,

$$(f_{L1}/f_{L2})^2 = I_{L2}/I_{L1} \tag{8}$$

By substituting the above equation for I_{L2} in terms of I_{L1}, we get

$$\begin{aligned} \varphi_{L1} - \varphi_{L2} &= (f_{L1}/f_{L2})^2 \times I_{L1} - I_{L1} + \lambda_{L1} N_{L1} - \lambda_{L2} N_{L2} \\ \varphi_{L1} - \varphi_{L2} &= \left((f_{L1}/f_{L2})^2 - 1\right) \times I_{L1} + \lambda_{L1} N_{L1} - \lambda_{L2} N_{L2} \end{aligned} \tag{9}$$

Simplifying Eq. (9) we get,

$$I_{L1} = 1/\left((f_{L1}/f_{L2})^2 - 1\right) \times (\varphi_{L1} - \varphi_{L2}) + 1/\left((f_{L1}/f_{L2})^2 - 1\right) \times (\lambda_{L2} N_{L2} - \lambda_{L1} N_{L1}) \tag{10}$$

Substituting the above expression for I_1 in Eq. (6) we get,

$$\begin{aligned} MP_{L1} - \lambda_{L1} N_{L1} = P_{L1} - \varphi_{L1} &- 2/\left((f_{L1}/f_{L2})^2 - 1\right) \times (\varphi_{L1} - \varphi_{L2}) \\ &+ 2/\left((f_{L1}/f_{L2})^2 - 1\right) \times (\lambda_{L2} N_{L2} - \lambda_{L1} N_{L1}) \end{aligned} \tag{11}$$

Rearranging the terms in Eq. (11) we get,

$$\begin{aligned} &MP_{L1} - \left(\lambda_{L1} N_{L1} - 2/\left((f_{L1}/f_{L2})^2 - 1\right) \times (\lambda_{L2} N_{L2} - \lambda_{L1} N_{L1})\right) \\ &\quad = P_{L1} - \left((f_{L1}/f_{L2})^2 + 1\right)/\left((f_{L1}/f_{L2})^2 - 1\right) \times \varphi_{L1} + 2/\left((f_{L1}/f_{L2})^2 - 1\right) \times \varphi_{L2} \end{aligned} \tag{12}$$

$$\begin{aligned} MP_{L1} = &\left(\lambda_{L1} N_{L1} - 2/\left((f_{L1}/f_{L2})^2 - 1\right) \times (\lambda_{L2} N_{L2} - \lambda_{L1} N_{L1})\right) P_{L1} \\ &- \left((f_{L1}/f_{L2})^2 + 1\right)/\left((f_{L1}/f_{L2})^2 - 1\right) \times \varphi_{L1} + 2/\left((f_{L1}/f_{L2})^2 - 1\right) \times \varphi_{L2} \end{aligned} \tag{13}$$

In above equation $\left(\lambda_{L1} N_{L1} - 2/\left((f_{L1}/f_{L2})^2 - 1\right) \times (\lambda_{L2} N_{L2} - \lambda_{L1} N_{L1})\right)$ is constant and expectation of MP_{L1} is assumed as zero. The impact of the multipath error and its variation with respect to elevation angle of the satellites for the entire duration of observation are analyzed. This analysis will be helpful in kinematic applications where multipath signal becomes more arbitrary, particularly in aircraft navigation and missile guidance where the reflecting geometry and the environment around the receiving antenna changes relatively in random way [29].

4. Results and discussion

Statistical analysis of the results shows that multipath error is too large to neglect. These errors are estimated for location having ECEF coordinates as x_u = 706970.90 m, y_u = 6035941.02 m, and z_u = 1930009.58 m in the Indian subcontinent for typical ephemerides data collected on March 11, 2019, from the dual frequency GPS receiver located at Department of Electronics and Communication Engineering, Andhra University College of Engineering, Visakhapatnam (Lat: 17.73°N/Long: 83.319°E), India. The city of Visakhapatnam with an area of 11,161 sq.kms is surrounded by Eastern Ghat Range, viz. Kailasa, Yarada and Narava hill ranges on north, south and west, respectively, and Bay of Bengal in the east. Due to this geography the GPS signals bound to get reflected. The data are collected on March 11, 2019, for entire 24 hrs with an epoch interval of 30 s. On March 11, 2019, the global geomagnetic activity index, i.e., Kp-index was 3. The average level for geomagnetic activity, i.e., Ap index was 2 and the noise level generated by the sun at a wavelength of 10.7 cm at the earth's orbit, i.e., solar index F10.7 was 69.5. These indices imply that on this particular day there was no solar storm or geomagnetic storm and the solar activity was normal. The solar activity affects the ionization in ionosphere and hence the signal propagation through this layer. The expression derived above for multipath error is ionospheric delay free. Hence the estimated multipath error is unaffected by ionospheric delay. For the observation period of 24 h, error analysis which is supported by the relevant graphs and tables are presented in this chapter. During this observation period, out of 32 satellites, minimum of 9 satellites were visible in each epoch. Though the error is computed and analyzed for all the visible satellites, the multipath error estimated for SV PRN07, 23, 28 and 31 are presented in this chapter. Navigation solution for each epoch is calculated using pseudoranges (multipath error corrected) of all the visible satellites.

Table 2 illustrates the multipath error for four satellites. Similar results were also obtained for all the visible satellites. Table also details the error in receiver position distance from the surveyed location. **Figure 2** shows the trajectories of the satellites 07, 23, 28 and 31 with respect to elevation and azimuth angles. The subplots of **Figure 3** show the change in multipath error with respect to change in elevation angle. **Figure 2** shows that the satellites were visible to the receiver at low elevation angle and rose to highest elevation angle of 70°, 84°, 52° and 82°, respectively. The receiver continued tracking the satellites. The satellites went out of the sight of the receiver when they set with low elevation angle. In each of the subplots **Figure 3** two curves of the change in multipath against the elevation angle are shown. One curve indicates the multipath error while the satellite was rising and the other when it was setting after reaching the highest elevation angle. From **Figure 3(a)–(d)**, it is

	Pseudorange multipath error on L1 frequency[m]				Error in receiver position distance [m]
	SV PRN07	*SV PRN23*	*SV PRN28*	*SV PRN31*	
Min	7.362	14.11	40.21	9.136	−25.8
Max	14.32	18.79	52.88	13.52	31.49
Standard deviation	1.816	1.439	1.984	1.019	10.78

Table 2.
Pseudorange multipath error for satellites signal on L1 frequency.

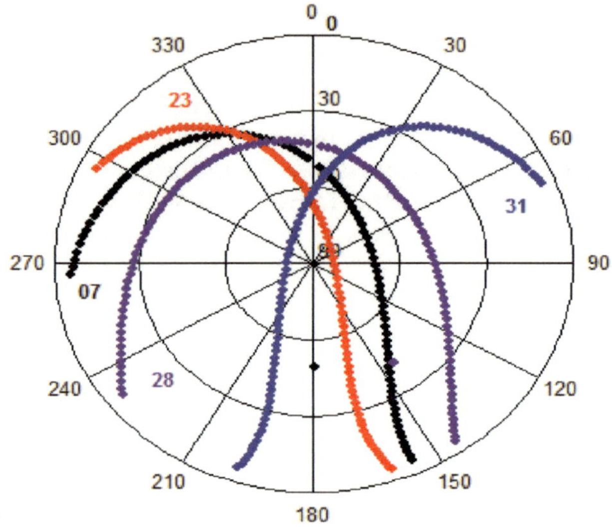

Figure 2.
Sky plot for the mentioned satellite orbits as viewed from GPS receiver located at Department of ECE, Andhra University (Lat: 17.73°N, Long: 83.31°E).

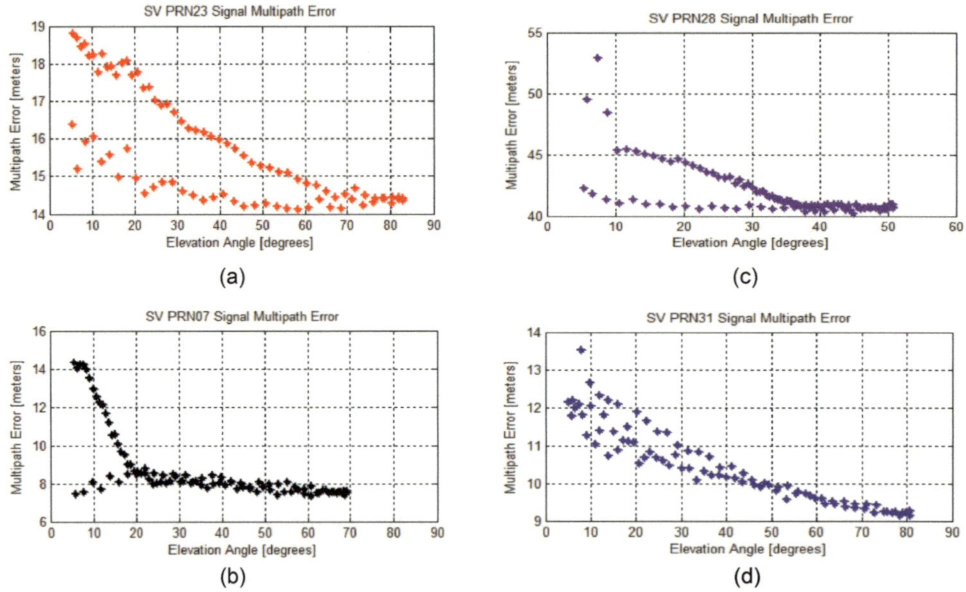

Figure 3.
(a)–(d) Pseudorange multipath error for respective satellites against the elevation angle for the observation period of 24 h.

observed that for the elevation angle of less than 10°, the multipath errors are 14.32 m, 18.79 m, 52.88 m and 13.52 m respectively. **Figure 4** shows the receiver position error in distance with respect to actual ECEF coordinates of the receiver. The figure shows the maximum error of 30 m. This is due to the residual errors in the pseudorange measurement. Though the pseudoranges of all the visible satellites are corrected for multipath error but the errors other than multipath remain uncorrected. The standard deviation of position error in distance is 10.78 m.

Figure 4.
Error in distance of GPS receiver position from surveyed position over the observation period of 24 h on March 11, 2019.

5. Conclusions

The statistics and result analysis comprises of investigation of error magnitude variations over a period of 24 h. Signals transmitted from the satellites, visible at low elevation angles, travel a longer path through the propagation medium and are subjected to multiple reflections than the satellites at higher elevation angle. From the results, the maximum multipath error of 52 m is observed for SVPRN28 at an elevation angle of 8° and minimum error of 7 m is observed for SVPRN7 at an elevation angle of 70°. But for SVPRN23, the minimum error is 14 m even though the elevation angle of the satellite is 84°. This is due to the multiple reflections the signal underwent for that particular azimuth angle, which determines the direction of the signal. The GPS receiver location is surrounded by high hill ranges and large water body, this would have led to multiple reflections and hence the large multipath error in spite of high elevation angle of the satellite. The receiver position error in distance with respect to actual position is 30 m. This is due to the residual errors in the pseudorange measurement as the errors other than multipath remain uncorrected. Along with the multipath error mitigation technique mentioned in this chapter, if other errors are also corrected the receiver position will be more accurate. The proposed algorithm to estimate multipath error is essential for all precise navigation applications (e.g., CAT I/II aircraft landings, missile navigation) and especially in surveying applications in urban canyon. The impact analysis done in this chapter will also be a valuable aid in selecting a location to set up the GPS receiver antenna with least multipath error for surveying, aircraft navigation and tracking.

Author details

Bharati Bidikar, Babji Prasad Chapa, Mogadala Vinod Kumar
and Gottapu Sasibhushana Rao*
Department of Electronics and Communications Engineering, Andhra University
College of Engineering (A), Andhra University, Visakhapatnam, India

*Address all correspondence to: sasigps@gmail.com

IntechOpen

© 2020 The Author(s). Licensee IntechOpen. This chapter is distributed under the terms of the Creative Commons Attribution License (http://creativecommons.org/licenses/by/3.0), which permits unrestricted use, distribution, and reproduction in any medium, provided the original work is properly cited.

References

[1] Bidikar B, Sasibhushana Rao G, Ganesh L, Santosh Kumar MNVS. GPS C/A code multipath error estimation for surveying applications in urban canyon. In: Microelectronics, Electromagnetics and Telecommunications. Lecture Notes in Electrical Engineering. Vol. 372. Springer. 2016. pp. 135-142. ISBN: 978-81-322-2728-1

[2] Sunehra D. Estimation of prominent global positioning system measurement errors for Gagan applications. European Scientific Journal. 2013;**9**(15). ISSN: 1857-7881 (Print) e-ISSN: 1857-7431 68

[3] Borre K, Strang G. Linear Algebra Geodesy and GPS. USA: Wellesley-Cambridge Press; 1997

[4] Demyanov VV, Sergeeva MA, Yasyukevich AS. GNSS high-rate data and efficiency of ionospheric scintillation indices. In: Demyanov V, editor. Ionospheric and Atmospheric Threats for GNSS and Satellite Telecommunications. Croatia. ISBN: 978-1-78985-996-6.: IntechOpen; 2019. DOI: 10.5772/intechopen.90078

[5] Nahavandchi H et al. Correlation analysis of multipath effects in GPS-code and carrier phase observations. Survey Review. 2010;**42**(316):193-206

[6] Shkarofsky IP et al. Multipath depolarization theory combining antenna with atmospheric and ground reflection effects. Annales des Télécommunications. 1981;**36**(1-2): 83-88

[7] Jayanta KR. Mitigation of GPS code and carrier phase multipath effects using a multi-antenna system [Thesis]. Calgary, Alberta: The University of Calgary; 2000

[8] Matera ER, Peña AJG, Julien O, Ekambi B. Characterization of pseudo range multipath errors in an urban environment. In: ITSNT 2018, International Technical Symposium on Navigation and Timing; October, Toulosue, France. 2018. DOI: 10.31701/itsnt2018.22.hal-01890371f

[9] Clare A, Lin T, Lachapelle G. Effect of GNSS receiver carrier phase tracking loops on earthquake monitoring performance. Advances in Space Research. 2017;**59**(11):2740-2749

[10] Cheng L, Chen J, Gan M. Multipath error analysis of carrier tracking loop in GPS receiver. In: Proceedings of the 29th Chinese Control Conference; Beijing. 2010. pp. 4137-4141

[11] Macabiau C, Roturier B, Benhallam A, Chatre E. Performance of GPS receivers with more than one multipath. In: ION GPS 1999, 12th International Technical Meeting of the Satellite Division of The Institute of Navigation, Nashville, United States. 1999. pp. 281-288. ffhal-01021687f

[12] Cui J, Kouguchi N. Ocean wave observation by GPS signal. In: OCEANS 2011 IEEE; Spain, Santander. 2011. pp. 1-7

[13] Avram A, Schwieger V, El Gemayel N. Experimental results of multipath behavior for GPS L1-L2 and Galileo E1-E5b in static and kinematic scenarios. Journal of Applied Geodesy. 2019;**13**(4):279-289

[14] Falkenberg W, Kielland P, Lachapelle G. GPS differential positioning technologies for hydrographic surveying. In: Proceedings of IEEE PLANS, Orlando, December. 1988. pp. 310-317

[15] Lachapelle G, Falkenberg W, Neufeldt D, Keilland P. Marine DGPS using code and carrier in multipath environment. In: Proceedings of ION GPS-89, Colorado, Springs, September 27–29. 1989. pp. 343-347

[16] Padma B, Kai B. Performance analysis of dual-frequency receiver using combinations of GPS L1, L5, and L2 civil signals. Journal of Geodesy. 2019;**93**:437-447. DOI: 10.1007/s00190-018-1172-9

[17] Axelrad P, Comp CJ, MacDoran PF. Use of signal-to-noise ratio for multipath error correction in GPS differential phase measurements. In: Proceedings of ION GPS-94; 20–23 September; Salt Lake City, USA. 1994. pp. 655-666

[18] Xia L, Liu J. Approach for multipath reduction using wavelet algorithm. In: Proceedings of ION GPS 2001; 11–14 September; Salt Lake City, USA. 2001. pp. 2134-2143

[19] de Souza EM, Monico JFG. Wavelet shrinkage: High frequency multipath reduction from GPS relative positioning. GPS Solutions. 2004;**8**:152-159

[20] Satirapod C, Ogaja C, Wang J, Rizos C. An approach to GPS analysis incorporating wavelet decomposition. Artificial Satellites. 2001;**36**:27-35

[21] Zheng DW, Zhong P, Ding XL, Chen W. Filtering GPS time-series using a Vondrak filter and cross-validation. Journal of Geodesy. 2005;**79**:363-369

[22] Ge L, Han S, Rizos C. Multipath mitigation of continuous GPS measurements using an adaptive filter. GPS Solutions. 2000;**4**:19-30

[23] Xie P et al. Measuring GNSS multipath distributions in urban canyon environments. IEEE Transactions on Instrumentation and Measurement. 2015;**64**(2)

[24] Kaplan ED. Understanding GPS: Principles and Applications. 2nd ed. Boston, USA: Artech House Publishers; 2006

[25] Townsend B, Fenton R. A practical approach to the reduction of pseudo-range multipath errors in an L1 GPS receiver. In: 7th Intenational Technical Meeting of the Satellite Division of the U.S. Inst. of Navigation, 20–23 September; Salt Lake City, Utah. 1994. pp. 143-148

[26] Rao GS. Global Navigation Satellite Systems. 1st ed. India: McGraw-Hill; 2010

[27] Pratap M, Per E. Global Positioning System: Signals, Measurements and Performance. 2nd ed. New York: Ganga-Jamuna Press; 2006

[28] Parkinson BW, Spilker JR. Global Positioning System: Theory and Applications. Washington DC: American Institute of Aeronautics and Astronautics; 1996

[29] Happel DA. Use of military GPS in a civil environment. In: Proceedings of the 59th Annual Meeting of the Institute of Navigation and CIGTF 22nd Guidance Test Symposium; Albuquerque, NM. 2003. pp. 57-64

Chapter 5

The Impact of Space Radiation Environment on Satellites Operation in Near-Earth Space

Victor U.J. Nwankwo, Nnamdi N. Jibiri and Michael T. Kio

Abstract

Energetic particles and electromagnetic radiation (EM) from solar events and galactic cosmic rays can bombard and interact with satellites' exposed surfaces, and sometimes possess enough energy to penetrate their surface. Among other known effects, the scenario can cause accelerated orbit decay due to atmospheric drag, sporadic and unexplainable errors in functions of sensitive parts, degradation of critical properties of structural materials, jeopardy of flight worthiness, transient and terminal health hazard to both onboard passengers and astronauts, and sometimes a catastrophic failure that can abruptly end satellite mission. The understanding of the dynamics of the space radiation environment and associated effects is critically important for satellites design and operation in ionospheric plasma environment, in which satellites are designed to function. In this chapter we review some satellite anomalies associated with the space radiation environment and conclude with mitigation effort that can reduce such impact.

Keywords: solar activity, energetic particles, radiation environment, single event effects, total ionizing dose, impact mitigation

1. Introduction

Solar activity drives dynamic changes in the atmosphere and ionosphere that can affect the performance and reliability of satellites in near-Earth space environment, as well as ground-based technological systems and services that rely on them. This condition is referred to as space weather. The principal medium through which the Sun's activity is communicated to the region of the near-Earth space environment, is the solar wind, which occurs in form of a continuous outflow of streams of energized charged particles and/or momentary eruption of large-scale, high-mass plasma known as coronal mass ejections (CMEs). Sources of energised particles and strong magnetic energy also include the solar flares and galactic cosmic ray, originating from outer space. The energetic particles and electromagnetic radiation from these processes form the near-Earth radiation environment and can be divided into (i) trapped radiation environment and (ii) transient radiation environments. The charged particles that are trapped or confined by the Earth's magnetic field to certain regions in space such as the Van Allen belts form the trapped radiation environment. The transient particles environment consists of energetic particles from solar events, and galactic cosmic radiation that exist in the interplanetary space regions and in the near-Earth regions. Satellites and other space application systems are vulnerable to

IntechOpen

both trapped and transient energetic particles since they are basically designed to operate in the space plasma environment. The particles can bombard and interact with satellites' surfaces, and sometimes posses enough energy to penetrate their exposed surfaces with possible access to their electrical, electronic and electrochemical components (EEECs). This scenario can induce sporadic and unexplainable errors in sensitive parts of spacecrafts, degrade the critical properties of their structural materials, jeopardize the flight worthiness of spacecrafts, constitute transient and terminal health hazard to both onboard passengers and astronauts, and even lead to total failure that can end the mission of affected spacecrafts [1, 2].

There are documented cases or evidence of satellites anomaly associated with space weather (or space radiation environment). In their study, Iucci et al. [3] verified and quantified the linkage between a large fraction of spacecraft anomalies and space weather perturbations. They compiled a large database of about 5700 anomalies registered by 220 satellites in different orbits over the period of 23 years (1971–1994). Their findings revealed that very intense fluxes (>1000 particles cm^{-2} s^{-1} sr^{-1} (pfu) at energy >10 MeV) of solar protons are linked to anomalies registered by the satellites in high-altitude (>15,000 km) near-polar (inclination >55°) orbits and to a much smaller extent to anomalies in geostationary orbits. They also reported that elevated fluxes of energetic (>2 MeV) electrons >10 8 cm^{-2} d^{-1} sr^{-1} are observed by the Geostationary Operational Environmental Satellites (GOES) on days with satellite anomalies occurring at geostationary and low-altitude (<1500 km) near-polar (>55°) orbits [3]. On the 22nd and 23rd of March 1991, an intense solar event occurred, which resulted to severe geomagnetic storms. This strong solar flare event with high energetic solar radiation caused disruption in high latitude point-to-point communication, and solar panel degradation on GOES-6 and -7 satellites, and was estimated to have decreased the expected lifetime of GOES-7 by 2 to 3 years. During the event, high energetic solar particles also increased the frequency of single event upsets (SEU) recorded by the spacecrafts; up to six geostationary satellites, including GOES-6 and -7, and the Tracking and Data Relay Satellite (TDRS)-1 had about 37 reported cases of SEU during the main phase of the event. SEU will be explained in detail in Section 3.2. Other impacts associated with this solar activity include the loss of automatic altitude control of the National Oceanic and Atmospheric Administration (NOAA)-11 satellite, increased satellite drag due to the heated atmosphere, which necessitated a massive update of the North American Air Defense Command (NORAD) catalogue of orbiting objects, and the complete failure of the geosynchronous orbiting Maritime European Communication Satellite (MARECS)-1 as a result of critical damage to its solar panels [4, 5].

On September 2009, South Africa's SumbandilaSat (in low Earth orbit [LEO]) was reported to have experienced a power distribution failure due to radiation shortly after its launch, which rendered the Z- and Y-axis wheel permanently inoperable. However, the satellite continued to work as a technology demonstrator until 25 August 2011 when it failed completely. Its failure was again attributed to solar storm event, which caused the satellite's onboard computer to stop responding to commands from the ground station [6]. On 5 April 2010, Galaxy 15 spacecraft (at geosynchronous altitudes) was reported to have experienced an anomaly that caused it to stop responding to any ground command [7]. The failure was attributed to an onboard electrostatic discharge (ESD), which led to a lockup of the field-programmable gate array within the spacecraft baseband communications unit. The interaction of the spacecraft with substorm-injected energetic particles caused the ESD after the spacecraft experienced surface and deep dielectric charging. A concise documentation of many other cases of satellite anomalies and losses that have been attributed to space weather can be found in several literatures (e.g., p. 33 of Refs. [8] and [9].

2. Solar activity and the space radiation environment

The Sun's activity varies with time and position on the Sun, and characterized by 11-year cycle, which can be divided into solar minimum and solar maximum phases. The sunspots (and other solar indices such as solar radio flux) are viewed as main indicators of solar activity cycle. They are transient phenomenon seen as dark patches against photospheric bright background on the Sun. Observations made over the past two centuries have shown that the number of sunspots vary periodically, moving from minimum to maximum count approximately every 11 years. **Figure 1** show a historic sunspot number. The latest solar cycle (cycle 24) peaked around year 2014. Currently, solar activity is on the decline and has been predicted to reach its minimum in late 2019 or 2020, while the solar maximum is expected to occur between 2023 and 2026 [10].

Solar energetic events such as high-speed solar wind streams (HSS), solar flares and CMEs that give rise to solar particle events and geomagnetic storms affecting the space environment are more frequent during solar maximum. Therefore, their impact on the atmosphere and air-based technology are expected to be higher during this phase of the solar cycle than the declining or minimum phase. Solar events and associated phenomena mainly contribute to trapped and transient energetic particles in near space that constitute the space radiation environment, in addition to galactic cosmic ray from outer space. The summary of types of space radiation, their origin or sources, and where they are important is shown in **Figure 2**.

2.1 The trapped particles

When charged particles from the solar wind encounters and interacts with the Earth's magnetic field, it compresses it sun-ward, forming the magnetosphere (see, **Figure 3**). This scenario creates a supersonic shock wave known as the Bow Shock. The solar wind drags out the night-side of the inner magnetosphere. This extension is known as the magnetotail. Although the magnetosphere is constantly being bombarded by charged particles, they are being deflected and cannot easily penetrate the region; however, some particles gain entrance through the polar region and become trapped in the Earth's magnetic field. The trapped particles are contained in one of two doughnut-shaped magnetic rings surrounding the Earth called the Van Allen radiation belts, **Figure 3**. The inner belt contains a fairly stable population of protons with energies exceeding 10 MeV. The outer belt contains mainly electrons

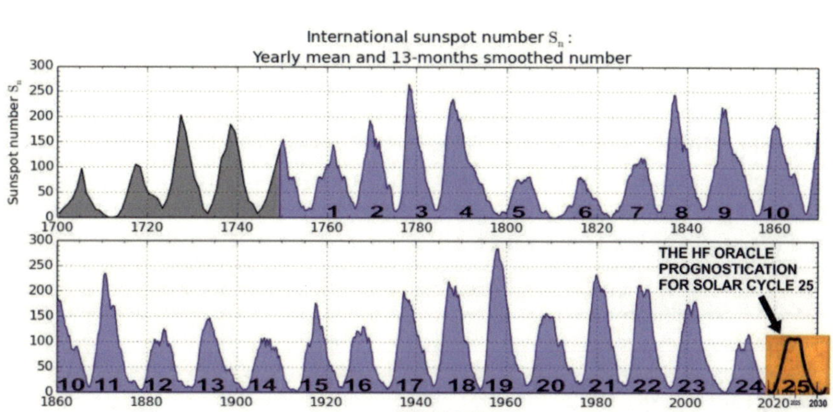

Figure 1.
Historic sunspot number (source: SILSO graphics (http://sidc.be/silso) Royal Observatory of Belgium).

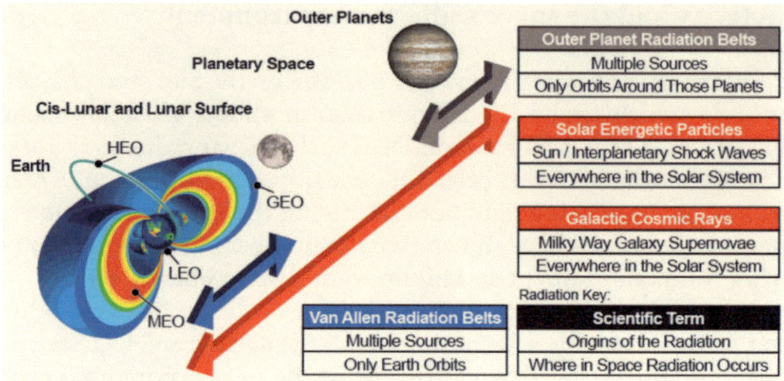

Figure 2.
Summary of types of space radiation, their origin or sources, and where they are important in the outer planets, planetary space and Earth, including the low Earth orbit (LEO), geostationary orbit (GEO), medium Earth orbit (MEO) and high Earth orbit (HEO) (source: Ref. [11]).

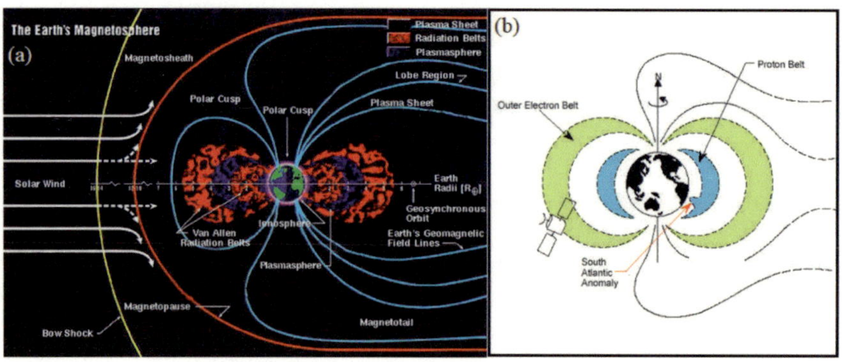

Figure 3.
(a) The Earth's magnetosphere showing the Van Allen radiation belt. (b) Outer and inner (proton) belt (source: Ref. [12]).

with energies up to 10 MeV. The charged particles which compose the belts circulate along the Earth's magnetic lines of force. These lines of force are known to extend from the area above the equator to the North Pole, to the South Pole, and then circle back to the Equator. There is a part of the inner Van Allen belt (VAB) that dips down to about 200 km into the upper region of the atmosphere over the southern Atlantic Ocean off the coast of Brazil. This region is known as the South Atlantic Anomaly (SAA). The dip results from the fact that the magnetic axis of the Earth is tilted approximately 11° from the spin axis, and the center of the magnetic field is offset from the geographical center of the Earth by 280 miles. The largest fraction of the radiation exposure received during spaceflight missions has resulted from passage through the SAA. Low inclination flights typically traverse a portion of the SAA up to six or seven times a day (see **Figure 3**).

2.2 The transient particles

The transient particles or radiation environments consist of particles from solar events such as solar wind, solar flares, CMEs and galactic cosmic radiation in the interplanetary and near-Earth space regions. The solar wind consists of relatively low energy electrons and protons that can significantly affect externally mounted spacecraft components. Solar flares are also a major contributor to the overall ionizing radiation level. A solar flare can emit and accelerate energetic particles or protons in the interplanetary space that can reach Earth within 30 minutes of the flare's peak. CMEs can propagate into the solar wind and drive shocks, which in turn accelerates solar energetic particles, and also deflect the galactic cosmic rays (GCRs) entering the heliosphere [13, 14]. CME can cause geomagnetic storms and other associated phenomena, leading to large-scale disturbances with adverse consequences in the geospace environment that can affect satellite systems.

2.3 Galactic cosmic radiation

Galactic cosmic radiations (GCR) are not directly connected to our Sun. They originate from outside the solar system. GCR consists of ionized atoms ranging from a single proton up to a uranium nucleus. The flux level of these particles is very low. Notwithstanding, they produce intense ionization as they pass through matter because they travel at a speed that is very close to that of light, and because some of them are composed of very heavy elements such as iron [15]. The energy of cosmic rays is usually measured in units of mega electron volt (MeV), or the giga electron volt (GeV). Most GCRs have energies between 100 MeV and 10 GeV. Cosmic rays include essentially all of the elements in the periodic table; about 85% protons, 14% alpha particles, and 1% heavy nuclei [16]. The Earth's magnetic field provides natural shielding from both cosmic and solar particles depending primarily on the inclination and secondarily on the altitude. As inclination reaches auroral to polar regions, a satellite is outside the protection of the geomagnetic field lines. At polar orbits intense fluxes of energetic electrons, known as precipitating electrons, propagate down along magnetic field lines (and create the aurora), and as altitude increases, the exposure to these particles gradually increases [12].

3. Effects of space particles and radiation environment on satellites operation in near-Earth space environment

When charged trapped or transient particles from solar events or cosmic sources bombards and interacts with the exposed surfaces of spacecraft, their effects can affect the system in a several ways. The effects from the natural space environment include spacecraft charging (SC), single event effects (SEEs), total ionizing dose (TID), and displacement damage (DD). However, the specific effect depends on the type of incident particle, its energy and probably the source. Trapped heavy ions do not have sufficient energy to generate the ionization required to cause SEEs, and they do not make a significant contribution to TID. Galactic cosmic rays and cosmic solar particles, which are heavily influenced by solar flares and trapped protons in the radiation belts, can cause SEEs, but electrons are not known to cause SEEs. Although their physical mechanisms are different, the ionizing radiation of the space environment causes both TID and SEEs. Charged particle effects in the space environment are summarized below according to the particle source.

3.1 Spacecraft charging (SC)

Spacecraft charging (SC) is the build-up of charge on spacecraft surfaces or in the spacecraft interior; SC causes variations in the electrostatic potential of a spacecraft surface with respect to the surrounding plasma environment, and potential variations in different portions of the spacecraft [17]. The major natural space environments which contribute to SC include the thermal plasma environment, high energy electrons, solar radiation and magnetic fields. Although SC has many effects, electrostatic discharges appear to be the most dangerous of all. Electrostatic discharges can cause structural damage, degradation of spacecraft components and operational anomalies due to damages to electronics. SC can be categorised into two: Surface charging which include differential charging, and internal dielectric charging. Surface charging is caused by low energy plasma (<100 keV) and photoelectric currents. Surface charging can either be absolute or differential. Absolute charging occurs when the satellite potential relative to the ambient plasma is charged uniformly, while differential charging occurs when parts of the spacecraft are charged to different potential relative to one another. Differential charging can also be caused by satellite self-shadowing. The charge control mechanism, and differential charging in spacecrafts are depicted in **Figure 4**. Differential charging of spacecraft surfaces is more detrimental than the absolute charging (relative to ambient plasma). The former can have a discharge effects that can disrupt satellite operations such as physical materials damage and electromagnetic interference (EMI) generation, and resultant transient pulses. Discharge consequences also include noise in data and wiring, sputtering and attraction of chemically active species [18]. Differential charging has been reported after geomagnetic sub-storms, which result in the injection of keV electrons into the magnetosphere.

Internal charging is caused by high-energy electrons (>100 keV), which penetrate into the spacecraft equipment where they deposit charge inside insulating materials [8]. Internal discharge is more damaging since it occurs within dielectric materials and well-insulated conductors, which are in close proximity to sensitive electronic circuitry [19]. Based on data from the Combined Release and Radiation Effects Satellite (CRRES) obtained at GEO, most environmentally induced spacecraft anomalies result from deep dielectric charging and the resulting discharge pulses and not from surface insulator charging or single-event upsets [20].

3.2 Single event effects (SEEs)

Single event effects (SEEs) are individual events which occur when a single incident ionizing particle deposits enough energy to cause an effect in a device. SEEs are generally caused by two space radiation sources: high energy protons, and cosmic rays. Single event phenomenon can be classified into four: (i) single event upset (SEU), (ii) single event latch-up (SEL), (iii) single event burnout (SEB) and (iv) single event gate rupture (SEGR). SEU is a change of state caused by ions or electromagnetic radiation striking a sensitive node in a micro-electronic device, such as in a microprocessor, semiconductor memory, or power transistors. The state change is a result of the free charge created by ionization in or close to an important node of a logic element (e.g., memory bit). The error in device output or operation

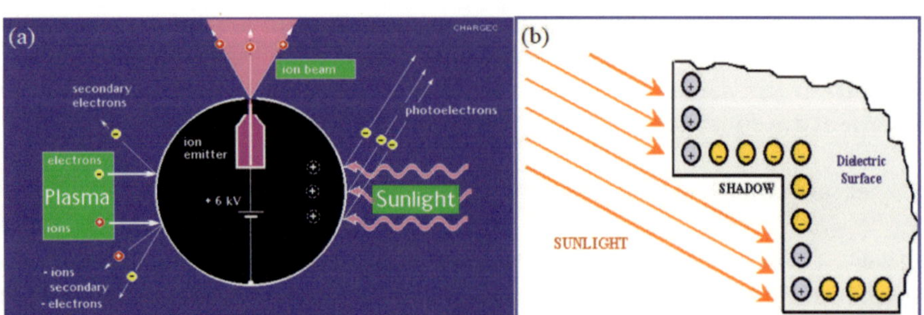

Figure 4.
(a) Satellite's charge control mechanism, and (b) differential charging in satellites due to self-shadowing (source: Ref. [12]).

caused as a result of the strike is called a soft error. The mechanisms for heavy ion and proton SEU in devices (e.g., dynamic random access memories (DRAM)), and galactic cosmic ray energy deposition in devices are depicted in **Figure 5**. SEU can cause a reset or re-writing in normal device such as in analogue, digital, or optical components, and may also have effects in surrounding interface circuitry. A severe SEU is the single-event functional interrupt (SEFI) in which an SEU in the device's control circuitry places the device into a test mode, halt, or undefined state. The SEFI halts normal operations, and requires a power reset to recover [1].

SEL is used in integrated circuits (ICs) to describe a particular type of short circuit which can occur in an improperly designed circuit. It is the generation of a low-impedance path between the power supply rails of a MOSFET circuit that can trigger a parasitic structure which disrupts proper functioning of the part and possibly even leading to its destruction due to over-current. SELs are hard errors, and can cause permanent damage. It can results in a high operating current, above device specifications, drag down the bus voltage, or damage the power supply. Latch-up can be caused by protons in very sensitive devices [22]. An SEL is corrected or cleared by a power off–on reset or power strobing of the device. SEL is strongly temperature dependent. If power is not removed quickly, catastrophic failure may occur due to excessive heating or metallization or bond wire failure [23].

SEB is a condition caused by high current state in a power transistor. It is a highly localized phenomenon, and includes burnout of the drain-source in power MOSFETs and BJTs, gate rupture, frozen bits, and noise in charged-coupled devices (CCDs). SEGR is the formation of a conducting path or localized dielectric breakdown in the gate oxide resulting in a destructive burnout. It occurs at MOSFETs, BJTs, and CMOS.

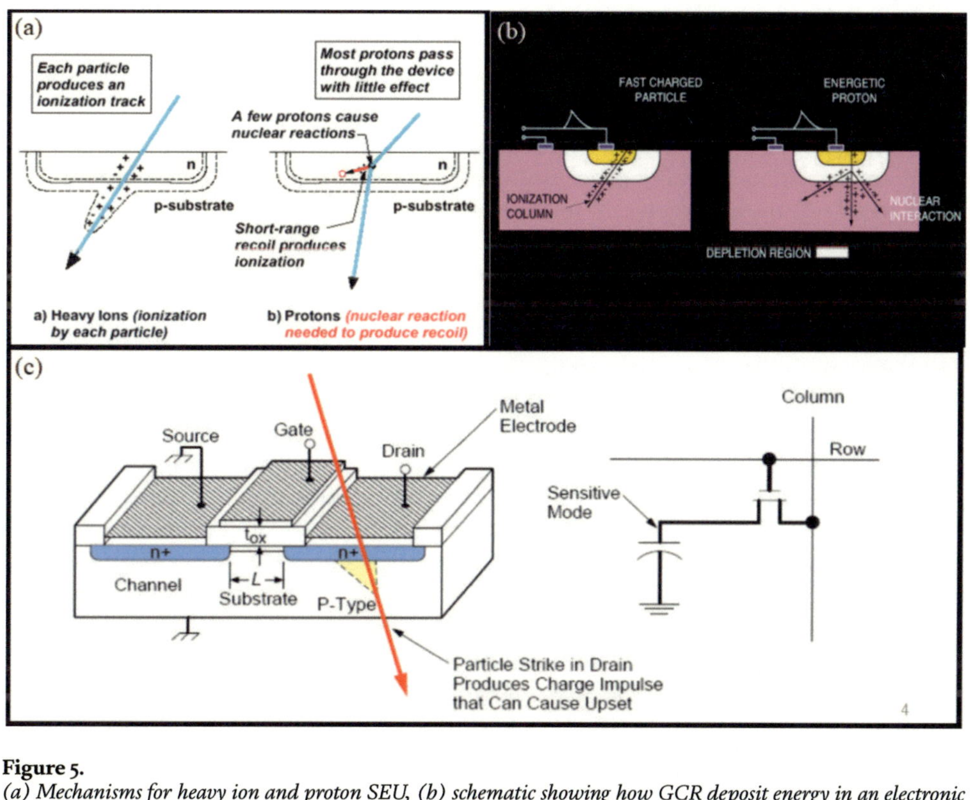

Figure 5.
(a) Mechanisms for heavy ion and proton SEU, (b) schematic showing how GCR deposit energy in an electronic device [12], and (c) upset mechanism for dynamic random-access memories (DRAMs) (from Ref. [21]).

Solar flare particle events pose the most extreme SEU producing environment, especially for spacecraft in interplanetary space [24]. Experiments aboard CRRES showed a significant increase during a solar flare [25]. Based on CRRES's data, most SEUs come from high energy protons through nuclear interactions and not through direct deposition from either protons or cosmic rays [20]. For LEO satellites, trapped protons, especially in the SAA, are the greatest SEE threat.

3.3 Total ionizing dose (TID)

Total ionizing dose (TID) refers to the amount of energy that ionization processes create and deposit in materials such as semiconductor or insulator when energized particles pass through it. TID can result in device failure or biological damage to astronauts. Radiation-induced trapped charges can build up in the gate oxide of a MOSFET and cause a shift in the threshold voltage. Such device cannot be turned off even at zero volts applied, if the shift is large enough. Under this condition the device is said to have failed by going into depletion mode [26]. TID is mostly due to electrons and protons, mainly from solar energetic particle event and passage through the SAA. In low Earth orbit, the main dose source is from electrons and inner belt protons, while the primary source is outer belt electron and solar protons in geostationary orbit. The first recorded satellite failure resulting from total dose was the Telstar. The satellite was launched a day after the Starfish nuclear test on 9 July 1962. The nuclear weapon of about 1.4 Megaton was detonated at an altitude of about 400 km above Johnston Island in the Pacific Ocean. The explosion produced beta particles (electrons) that were injected into the Earth's magnetic field, forming an artificial radiation belt. This artificial electron belt lasted until the early 1970s. Consequently, Telstar experienced a total dose 100 times that expected before its total failure. Up to seven satellites were destroyed by the Starfish nuclear test within 7 months mainly from solar cell damage [12].

3.4 Displacement damage (DD)

When energetic particles are incident on a solid material, they lose their energy to ionizing and non-ionizing processes as they travel through the material. The consequence of the energy loss is in the production of electron–hole pairs and atoms displacement or displacement damage. Vacancies (i.e., absence of an atom from its normal lattice position) and interstitials (i.e., movement of displaced atom into a non-lattice position) are the primary lattice defects that are initially created. The combination of a vacancy and an adjacent interstitial is known as a Frenkel or close pair. Two adjacent vacancies can form a defect known as divacancy. Also, larger local groupings of vacancies may occur in irradiated silicon. A defect resulting from vacancy and interstitials being adjacent to impurity atoms is known as defect-impurity complexes. Once formed by incident radiation, the defects will reorder to form more stable configuration. The extent to which defects alter the properties of bulk semiconductor material and devices depends on nature of the particular defects and the time following the creation of defect at a given temperature.

The effectiveness of radiation-induced displacement damage depends on factors such as bombardment condition, particle type and energy, irradiation and measurement temperature, time after irradiation, thermal history after irradiation, injection level, material type, impurity type and concentration [27]. Displacement damage causes degradation of materials and device properties. **Figure 6** depicts the collision between an incoming particle and a lattice atom, causing the displacement

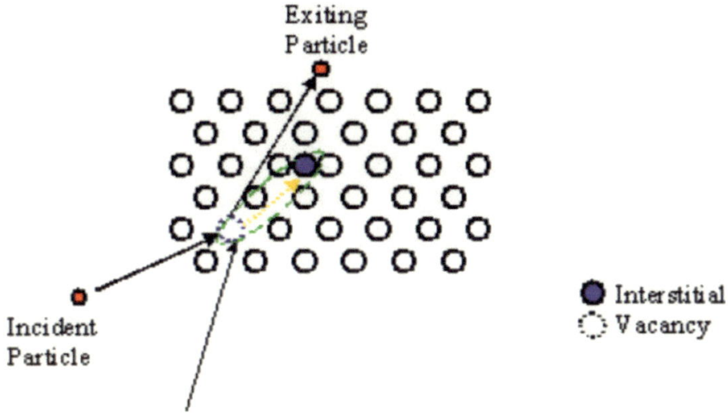

Figure 6.
Displacement of atom from its original lattice position by incoming particle through collision [12].

of the atom from its original lattice position. Displacement damage can also degrade minority carrier lifetime, and a typical effect would be degradation of gain and leakage current in bipolar transistors [12].

4. The stopping power, range of particles and deposited dose in spacecraft materials

The review presented here include portion of the work [1], part of which was published in [2]. We analyzed particles, electrons and protons flux of various energies from NOAA database for 3 months (April–June 2010). The mass stopping power, range and possible deposited dose of protons were calculated, and applied to the scenario of possible interaction of the particles with satellite surface and its electrical, electronic and electrochemical components.

4.1 Stopping power

Stopping power is the average energy loss of a particle per unit length (measured in MeV/cm) when passing through the material. Charged particles are known to ionize the atom or molecule which they encounter when passing through matter, and they lose energy in the process. The stopping power depends on the type and energy of the particle and on the properties of the material it passes. Although numerical values and units are identical for both quantities, the Stopping power refers to the property of the material while energy loss per unit path length describes what happens to the particle. The density of ionization along the particles path is proportional to the stopping power of the material because the production of an ion pairs requires a fixed amount of energy [28]. The Bethe-Bloch formula for stopping power derived from relativistic quantum mechanics is given by:

$$S = -\frac{dE}{dx} = \frac{4\pi z^2 k_0^2 e^4}{m v^2}\left[\ln\frac{2mv^2}{I} - ln\left(1 - \frac{v^2}{c^2}\right) - \frac{v^2}{c^2}\right]$$

where z is the atomic number of the heavy particle, e is magnitude of the electron charge, m is the electron rest mass, c is the speed of light, I is the mean excitation energy of the medium, v is the velocity of the particle and k_0 is the Boltzmann constant ($= 8.99 \times 10^9$ N m^2 C^{-2}).

The mass stopping power of the material is obtained by dividing the stopping power by the density (ρ) of the material. It is a useful quantity because it expresses the rate of energy loss of the charged particle per g/cm^2 of the medium traversed [28].

$$S = -\frac{dE}{\rho dx}$$

4.2 The range of particle

The range R of a particle (e.g., proton) of initial kinetic energy E_k and mass m is the mean distance it travels before coming to a stop. R depends on the particle type, initial energy and the material through which it traverses. A theoretical approach to the determination of charged particle range utilizes stopping power expression. The range of a proton computed by numerical integration of the stopping power using the continuous slowing down approximation (CSDA) is given by:

$$R = \int_{E_{min}}^{E_{max}} \left(-\frac{dE}{\rho dx}\right)^{-1} dE + R(E_{min})$$

where $R\ (E_{min})$ is the measured range at minimum energy E_{min} which is added to the integral equation and treated as a constant for a particle and material. For the calculations of ranges for proton E_{min} is taken to be as 1 MeV as much data is available at 1 Mev. $R\ (E_{max})$ is the measured range at maximum energy E_{max}.

In previous work we used the empirical relations suggested by [28] to calculate the mass stopping power of particles in spacecraft materials [1, 2]. However, we anticipate limitations in the equations because they were originally formulated for low energy particles. Values obtained using Bethe's equations are higher and assumed more accurate at higher particle energies.

4.3 Dose deposition and absorption

The total ionizing dose (TID), explained in Section 3.3, can be measured in terms of the absorbed dose; which is a measure of the energy absorbed by matter. Absorbed dose is quantified using either a unit called the rad (radiation absorbed dose) or the SI unit which is the gray (Gy). 1 Gy = 100 rads = 1 J/kg. The total accumulated dose on a satellite depends on orbit altitude, orientation, and time spent in orbit. To compute TID we need to know the integrated particle energy spectrum, $\varnothing(E)$ or the fluence as a function of particle energy. The dose is a function of the particle flux. It becomes important as the spacecraft spends more time in the space radiation environment. The stopping power is used to determine dose from charged particle by the following relationship:

$$D = \varnothing \frac{dE}{\rho dx}$$

where $\varnothing$ is the particle fluence (i.e., the number of particles striking the material over a specified time interval).

Satellite and space probes typically encounter TID between 10 krad (100 Gy) and 100 krad(Si) (1000 Gy(Si)). The time taken, t (in years) for a satellite's component to fail due to total ionizing dose can be obtained by dividing the maximum absorbed dose or TID threshold by the total absorbed dose per year, given as:

$$t\ (in\ yrs) = \frac{TID_{threshold}}{Dose/yr}$$

We performed theoretical calculations to predict the mean time to failure of a model satellite due to TID. The assumption is that the model satellite's body is made of aluminum alloy and 20 mm thickness (without impact mitigation such as protective coating on the satellite), in which the electrical, electronic and electrochemical components (mainly of silicon (Si) and germanium (Ge) materials) are housed [1, 2]. Our calculations were based on particles with E $\geq$ 78 MeV. When particles of this energy range bombard and penetrate the satellite, parts of their energies are lost due to the stopping power of the alloy but the reminder constitute significant dose to the components. With continuous exposure, the dose continues to build over time until the threshold is exceeded leading to completed failure of the affected satellite. Our calculations showed that a dose of 10 krad can build up on the model satellite's component within 3 years and 100 krad within 29 years.

5. Mitigating the impact of space radiation environment

5.1 Spacecraft charging

The electrons impinging on spacecraft surface in the space environment are faster than their ion counterpart because of their very small mass (when compared to that of ions). As a result the ambient electron flux is usually more than the ambient ion flux, leading to high level negative charging of the spacecraft. The regions of concern (in space) for internal charging of spacecrafts is illustrated in [29] and shown in **Figure 7**. Spacecraft charging can be mitigated by the methods of electron emission and ion reception [30]. Electron emission is the method in which a device pulls (or draws) electrons from the spacecraft ground and ejects them into space, while the ion reception is the method in which positive ions arrive at a spacecraft that is negatively charged to neutralize the negative charges. The former method is effective for reducing the negative charge of the spacecraft ground but not effective for dielectric surfaces. As a demerit, the process can lead to differential charging between the dielectric and the conducting ground. The later method is effective for mitigating negatively charged surface (whether dielectric or conductor), and reducing differential charging. However, it has the disadvantage of electroplating the entire spacecraft with extended use. Because each method has advantage (or disadvantage) over the other, the use of a combination of both types has been recommended. Other mitigation methods include plasma emission, partially conducting paint, polar molecule emission, mirror reflection and violet irradiation [31].

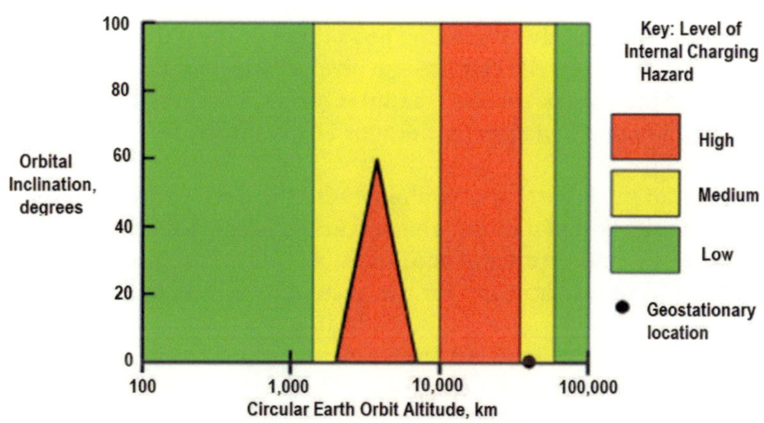

Figure 7.
Regions of concern for internal charging of spacecrafts in space (source: Ref. [29]).

5.2 Single event effects

For memories and data related devices, some of the error mitigation approach or methods include Parity check, cyclic-redundancy check (CRC) coding, Hamming code, Reed-Solomon (R-S) coding, convolutional encoding and overlying protocol (see: Ref. [32] and references therein). Parity is a single bit added to the end of a data structure, such that it states whether an odd or even number of 'ones' was in the structure. The parity method counts the number of logic-one states or 'ones' that are occurring in a data path. The CRC coding method detects if any errors occurred in a given data structure based on performing modulo-two arithmetic operations on a given stream of data, and interpreting the results as a polynomial. The hamming code method detects the position of a single error and the existence of more than one error in a data structure. The R-S code can detect and correct multiple and consecutive errors in a data structure. The convolutional encoding can also detect and correct multiple bit errors. However, it is distinguishable from block coding (e.g., R-S code) by interleave of the overhead or check bits into the actual stream of data instead of being grouped into separate words at the end of the data structure. Errors in the control-related devices can also be mitigated using some of the above mentioned methods. A more effective mitigation approach for control-related devices with complex difficulties (e.g., large scale integration circuitry or microprocessors) is the software-based mitigation, which includes tasks or subroutines dubbed health and safety (H&S). The H&S tasks can perform memory scrubbing that utilizes parity or other method on either external memory devices or registers that are internal to the microprocessor. In the software mitigation methods, the internal microprocessor timers can also be used to operate a watchdog timer or for passing H&S messages between spacecraft systems (see: Ref. [32] for more detail).

5.3 Total ionizing dose

TID on satellites system can be mitigated by methods such as shielding, derating and conservative circuit design [33]. Shielding is the processes of protecting spacecraft (and the occupants) from ionizing radiation using a configuration of appropriate massive materials. Derating refers to techniques usually employed in electrical power and electronic devices in which devices are operated at maximum power dissipation that is less than their rated value, with consideration of the case or body temperature, ambient temperature and the type of cooling mechanism used. This method can increase the safety margin between part design limits and applied stresses, consequently enhancing protection of the part [34]. Hardening of critical components in satellites at design level is also a viable method. This has, however, been the practice of satellite manufacturers. These methods can also be used to mitigate *Displacement damage* because DD is similar to TID as the effect is also cumulative [33].

Other important mitigation approach includes the development of appropriate environmental model that can mimic the perturbed scenarios that are expected under extreme space environmental condition. A well-accomplished or more sophisticated model should account for the individual effects of various solar forcing mechanisms, which cause fluctuations in neutral and ionized density [35]. One other very important mitigation approach to consider is the development of extensive warning system for solar energetic events. Although solar activity can be predicted days in advance but ascertaining their level of impact on the satellite and the Earth environment is quite challenging. Therefore, effective monitoring of solar activity is essential in order to be able to predict atmospheric or ionospheric

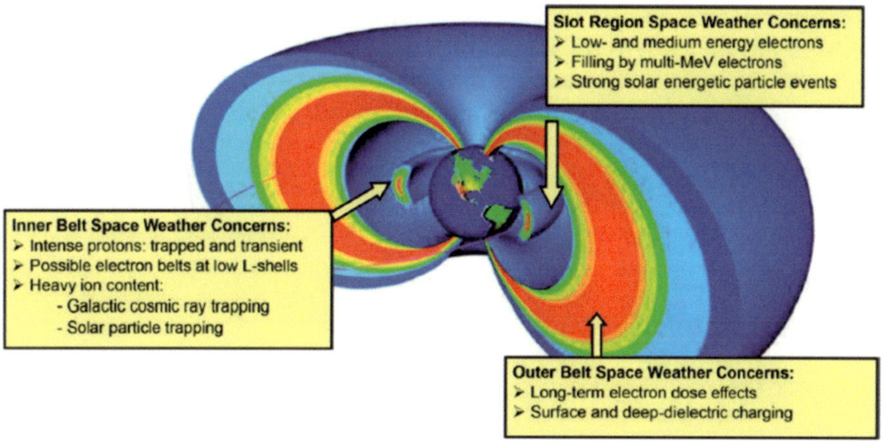

Figure 8.
Schematic diagram of Earth's radiation belts and their space weather concerns (from Ref. [36]).

responses to solar events and their consequence on satellite in orbit. In all, orbit consideration (and satellite's trajectory) is also important. Satellites in medium Earth orbit (MEO) and geostationary orbit (GEO) are subject to impacts of outer Van Allen radiation belt. LEO satellites encounter the most intense particle fluxes in the SAA [36], which is considered to be the main region where spacecrafts receive the largest fraction of the radiation exposure during spaceflight missions. The schematic diagram of Earth's radiation belts and their space weather concerns is shown in **Figure 8**.

6. Conclusion

The space radiation environment driven by solar activity (and galactic cosmic rays) poses potent and unequivocal treat to satellites in near-Earth space. Understanding atmospheric and ionospheric dynamic responses to solar-driven particles and radiation, and their space weather implications are critical and of practical importance to satellites design and operation. The specific effects of radiation environment on as satellite depends on the source, type and energy of incident particle, as well as the satellite's orbit and/or position at the time of solar energetic events. Radiation mitigation measures can increase the safety margin between part design limits and the applied stresses resulting from particles impact, consequently enhancing protection of the part. However, it is important that the solar maximum phase be given more consideration in all mitigation effort because the rate of impact is higher during this interval. Severe solar storms can occur during the solar maximum that can produce huge short-lived increase in radiation levels, as well as high levels of SEEs that current mitigation measures might not be able to bear [37]. Also as dependence on satellites services increase, the economic and societal risk associated with space weather also increases, and likely impact can be unprecedented. In view of this, a contingency plans that include the possibility of switching to or benefitting from other independent satellite services have been recommended [8]. The upcoming multi-constellation GNSS receivers can play a significant role in this regard, such that the individual GNSS receivers will be inherently robust to a satellite service denial. Space weather-induced enhancement of atmospheric drag on satellites and consequent accelerated orbit decay is also a major perturbing force to reckon with, for satellites in low Earth orbit [35, 38–42]. A concise review of the impact and mitigation of this phenomenon will be published in the future. We note that this review (on the space radiation effects on satellites and their mitigation methods) is succinct when compared to the large body of work in the subject area. Therefore, we encourage readers to also consult other well-accomplished texts for specific space radiation effect and the appropriate mitigation approach.

Author details

Victor U. J. Nwankwo[1*], Nnamdi N. Jibiri[2] and Michael T. Kio[3]

1 Space, Atmospheric Physics and Radio Wave Propagation Laboratory, Department of Physics, Anchor University, Lagos, Nigeria

2 Radiation and Health Physics Laboratory, Department of Physics, University of Ibadan, Ibadan, Nigeria

3 Engineering and Space Systems Department, National Space and Research Development Agency, Abuja, Nigeria

*Address all correspondence to: vnwankwo@aul.edu.ng

IntechOpen

© 2020 The Author(s). Licensee IntechOpen. This chapter is distributed under the terms of the Creative Commons Attribution License (http://creativecommons.org/licenses/by/3.0), which permits unrestricted use, distribution, and reproduction in any medium, provided the original work is properly cited.

References

[1] Nwankwo VJN. Determination of the stopping power and failure time of spacecraft components due to proton (e+) interaction using GOES 11 acquisition data [M.Sc. thesis]. Ibadan, Nigeria: University of Ibadan; 2010

[2] Jibiri NN, Nwankwo VUJ, Kio M. Determination of the stopping power and failure time of spacecraft components due to proton interaction using GOES 11 acquisition data. International Journal of Engineering, Science and Technology. 2011;**3**:6532-6542

[3] Iucci N, Levitin AE, Belov AV, Eroshenko EA, Ptitsyna NG, Villoresi G, et al. Space weather conditions and spacecraft anomalies in different orbits. Space Weather. 2005;**3**:1-16. S01001

[4] Shea MA, Smart DF, Allen JH, Wilkinson DL. Spacecraft problems in association with episodes of intense solar activity and related terrestrial phenomena during March 1991. IEEE Transactions on Nuclear Science. 1992;**39**:1754-1760

[5] Bedingfield KL, Leach RD, Alexander MB, editors. Spacecraft System Failures and Anomalies Attributed to Natural Space Environment. NASA REF-1390. NASA-MSFC, Alabama 35812; 1996

[6] Martin G. SumbandilaSat beyond repair. Defence Web. 2012. Available from: https://www.defenceweb.co.za/joint/science-a-defence-technology/sumbandilasat-beyond-repair/ [Retreived: 20 May 2019]

[7] Loto'aniu TM, Singer HJ, Rodriguez JV, Green J, Denig W, Biesecker D, et al. Space weather conditions during the Galaxy 15 spacecraft anomaly. Space Weather. 2015;**13**:484-502

[8] RAE. Extreme Space Weather: Impacts on Engineered Systems and Infrastructure. London: Royal Academy of Engineering (RAE); 2013

[9] Allen JH, Wilkinson D. Spacecraft charging: Then and now. In: Spacecraft charging technology conference, Albuquerque, NM, 20-24 September 2010

[10] National Weather Service (NWS). Solar experts predict the Sun's activity in Solar Cycle 25 to be below average, similar to Solar Cycle 24. National Oceanic and Atmospheric Administration (NOAA). Online Resource; 2019

[11] The National Academies of Sciences, Engineering and Medicine. Testing at the speed of light: The State of U.S. In: Electronic Parts Space Radiation Testing Infrastructure. Washington, DC: The National Academic Press; 2018. DOI: 10.17226/24993

[12] Holbert KE. Space Radiation Environmental Effects. Courses in Electrical Engineering, Arizona State University. 2007. Available from: http://holbert.faculty.asu.edu/eee560/spacerad.html [Retrieved: December 2010]

[13] Prolss GW. Physics of the Earth's Space Environment. Berlin, Heidelberg, Germany: Springer; 2004

[14] Gopalswamy N. Coronal mass ejections and space weather. In: Tsuda T, Fuji R, Shibata K, Geller MA, editors. Climate and Weather of the Sun-Earth System (CAWSES): Selected Paper from the 2007 Kyoto Symposium; 2009. pp. 77-120

[15] Mewalt RA. Cosmic Ray. 1996. Available from: http://www.srl.caltech.edu/personnel/dick/cos_encyc.html [Retrieved: July 2019]

[16] Adams L. Space Radiation Effects in Electronic Components. PA and

Safety Office, Brunel University. 2003. Available from: http://paso.esa.int/5_training.../training_03_space%20 radiation.ppt [Retrieved: March 2010]

[17] Mikaelian T. Spacecraft Charging and Hazards to Electronics in Space. York University Publication; 2001. arXiv:0906.3884; 2009

[18] Shaw RR, Nanevicz JE, Adamo RC. Observation of electrical discharges caused by differential satellite charging by magnetospheric plasmas. In: Rosen A, editor. Spacecraft Charging by Magnetospheric Plasma. Progress in Astronautics and Aeronautics. Vol. 47. 1976. pp. 61-76

[19] Leach RD, Alexander MB. Failure and Anomalies Attributed to Spacecraft Charging. AL: NASA RP-1375 Marshall Space Flight Centre; 1995

[20] Gussenhoven MS, Mullen EG, Brautigam DH. Improved understanding of the Earth's radiation belts from the CRRES satellite. IEEE Transactions on Nuclear Science. 1996;**43**(2):353-368

[21] Sayyah R, Macleod TC, Ho FD. Radiation-hardened electronics and ferroelectric memory for space flight systems. Ferroelectrics. 2011;**413**:170-175

[22] Nichols DK, Coss JR, Watson RK, Schwartz HR, Pease RL. An observation of proton-induced latchup. IEEE Transactions on Nuclear Science. 1992;**39**(6):1654-1656

[23] Mouret I, Allenspach M, Schrimpf RD, Brews JR, Galloway KF, Calvel P. Temperature and angular dependence of substrate response in SEGR. IEEE Transactions on Nuclear Science. 1994;**41**(6):2216-2221

[24] Adams JH, Gelman A. The effects of solar flares on single event upset rates. IEEE Transactions on Nuclear Science. 1984;**39**(6):1212-1216

[25] Campbell AB. SEU flight data from CRRES MEP. IEEE Transactions on Nuclear Science. 1991;**38**(6):1647-1654

[26] Oldham TR, McLean FB. Total ionizing dose in MOS oxides and devices. IEEE Transactions on Nuclear Science. 2003;**50**(3)

[27] Srour JR, Marshall CJ, Marshall PW. Review of displacement damage effects in silicon devices. IEEE Transactions on Nuclear Science. 2003;**50**:653-670

[28] Getachew A. Stopping Power and Range of Protons of Various Energies in Different Materials [M.Sc. thesis]. Ethiopia: Department of Physics, Addis Ababa University; 2007

[29] Garrett HB. Space Weather Impacts on Spacecrafts and Mitigation Strategies. California, USA, Pasadena, CA, USA: Jet Propulsion Laboratory, California Institute of Technology; 2012

[30] Lai ST. A critical overview on spacecraft charging control method. In: 6th Spacecraft Charging Technology Conference, AFRL-VS-TR-20001578; 2000

[31] Lai ST, Cahoy K. Spacecraft charging. In: Encyclopedia of Plasma Technology. Taylor & Francis; 2017. DOI: 10.1081/E-EPLT-120053644

[32] Maurer RH, Fraeman ME, Martin MN, Roth DR. Harsh environments: Space radiation environment, effects and mitigation. Johns Hopkins APL Technical Digest. 2008;**28**:17-29

[33] Shepherd SM. Spacecraft shielding. Online lecture series, Thayer Scholl of Engineering at Dartmouth College [Retrieved: October 2019]

[34] ReliaSoft Corporation. Reliability Basics. Reliability HotWire e-Magazine. HBM Prenscia Inc. Weibull.com; 2008

[35] Nwankwo VUJ, Chakrabarti SK, Weigel RS. Effects of plasma drag on low earth orbiting satellites due to solar forcing induced perturbations and heating. Advances in Space Research. 2015;**56**:47-56

[36] Baker DN, Erickson PJ, Fennell JF, Foster JC, Jaynes AN, Verronen PT. Space weather effects in the Earth's radiation belts. Space Science Reviews. 2017;**214**:17

[37] Hapgood and Thomson (for Lloyd's 360° Risk Insight). Space Weather: It's Impact on Earth and Implications for Business. London: RAL Space; 2010

[38] Nwankwo VUJ, Chakrabarti KS. Theoretical modeling of drag force impact on a model international space station (ISS) during variation of solar activity. Transactions of JSASS, Aerospace Technology Japan. 2014;**12**:47-53

[39] Nwankwo VUJ, Chakrabarti SK. Analysis of planetary and solar induced perturbations on trans-Martian trajectory of Mars missions before Mars orbit insertion. Indian Journal of Physics. 2015;**89**:1235-1245

[40] Nwankwo VUJ, Chakrabarti S. Effects of space weather on the ionosphere and LEO satellites orbital trajectory in the equatorial, low and mid-latitude regions. Advances in Space Research. 2018;**61**:1880-1889

[41] Nwankwo VUJ. Space weather: Responses of the atmosphere to solar activity and its implications for LEO satellites aerodynamic drag. In: Mukhopadhyay B, Sasmal S, editors. Exploring the Universe: From Near Space to Extra-Galactic. Springer International Publishing, Springer Nature Switzerland AG; 2018

[42] Nwankwo VUJ. Effects of space weather on Earth's ionosphere and nominal LEO satellites' aerodynamic drag [PhD thesis]. Kolkata, India: University of Calcutta; 2016

Chapter 6

Ionospheric Scintillation Modeling Needs and Tricks

Shishir Priyadarshi

Abstract

The wavelength of the radio-wave satellite signal is of the order of the minimal small-scale ionospheric irregularities (i.e., a few centimeters). As the satellite signal passes through the ionosphere, its interaction with the ionospheric irregularity structures causes refraction, reflection, and polarization in the satellite signal. Ionospheric irregularities degrade the trans-ionospheric radio-wave signal quality, between the satellite and the receivers, due to scintillation. The physics-based model often fails to produce global morphology during the extreme solar events, whereas empirical models based on the ionospheric scintillation data demonstrate better quality to forecast the scintillation effects during extreme solar event. It is really tricky to make a scintillation model that is sensitive to low and high solar activities as well as extreme solar events simultaneously. In the presented book chapter, we will discuss/review the needs and tricks of modeling ionospheric scintillation during extreme solar events as well as all weather and latitudinal cases. There are several aspects that influence the scintillation occurrence, its strength, and global distribution. The latitudinal dependence, local weather, solar/geomagnetic activity conditions, and local times are the widely accepted factors that control and influence ionospheric scintillation most. This book chapter discusses all these aspects and also suggests the ways to cast aside those factors that led to the wrong measure of scintillation indices.

Keywords: ionospheric scintillation, empirical scintillation model, GPS/GNSS scintillation modeling, global scintillation model

1. Introduction

Whenever the radio-wave signals pass through the ionospheric irregularities, these signals feel reflection, refraction, and scintillations (i.e., sharp and rapid carrier-phase variations and signal-to-noise ratio fading). In general it is more typical when the carrier-phase loss of lock happens due to the sharp signal-to-noise ratio fading ([1, 2] and references therein). As the wave propagates to the ground, scintillation models are needed to produce global as well as local ionospheric scintillation data and maps during the required solar activity condition, day, season, and geographic locations. Generally, scintillation models are made to serve special needs and not for all the spatial and temporal situations; due to this generally each scintillation model is not fit for all the geographic and solar activity conditions.

In general, scintillation models often have two significant limitations; firstly physics-based models often fail in producing scintillation morphology during

 IntechOpen

extreme solar and geomagnetic events. The second limitation is there is no possibility of a correction in the model once its algorithm is derived. In the presented book chapter, scintillation empirical modeling methods and their limitations are discussed. Geometrical effects contaminate scintillation observations most. We have discussed in this chapter how to overcome scintillation modeling limitations and use some tricks that replicate the actual scintillation morphology. The presented text in this chapter will enrich the knowledge of ionospheric modelers and make them understand how beginners should proceed with the ionospheric scintillation or ionospheric electron density as a model input data, if they decided to model certain ionospheric parameters (such as spectral index, turbulence strength parameters of the ionospheric irregularities, amplitude and phase fluctuations, scintillation indices, etc.).

2. Scintillation modeling tricks and correction for the geometry of propagation

As the elevation angle of the signal wave source changes, we observe the changes in the intensity of ionospheric scintillation. Such changes are significant as they are caused due to enhancement in the path of the radio-wave signal through the ionospheric irregularity layer with decreasing elevation angle and vice versa [3]. The signal ray path through the ionospheric irregularity also depends on the size and orientation of the ionospheric irregularity structure along with the height and thickness of the ionospheric irregularity layer [1]. All these elements influence the intensity of the ionospheric scintillation, and the effect caused by them on the trans-ionospheric radio-wave is termed as the geometrical effect [3–5]. In this section, we will cover the scintillation modeling tricks and method to minimize the multipath effect and impact ways to overcome other geometrical effects such as irregularity orientation with respect to the local geomagnetic field lines, multipath at higher elevation angle, radio-wave signals' extended path near the Earth's horizon, etc.

2.1 Scintillation modeling tricks

For the empirical scintillation modeling, first, we should need to have data for all the seasons and solar as well as geomagnetic activity situation. For instance, if we are planning to model ionospheric scintillation during high solar activity in winter months, we must have winter-month data collected during high solar activity period. Similarly, for the low solar activity period and in particular month scintillation modeling, we must have alike data. An adequate empirical modeling study expands understanding of ionospheric irregularities and can be used to better evaluate the impact of scintillation on different resources. The physics-based model often fails to produce global morphology during extreme solar events, whereas empirical models based on the ionospheric scintillation data demonstrate better scintillation effects during extreme solar events. It is really tricky to make a scintillation model which is sensitive to low and high solar activities as well as extreme solar events simultaneously.

The modeler should also derive the relationship between solar activity and provisional activity indices with the ionospheric scintillation data in order to make the scintillation data sensitive to the solar activity condition. One should always keep in mind that all the solar activity, geomagnetic, and provisional indices are not equally beneficial to all the geographic locations. Therefore, one should use the proper geo, solar, and provisional activity indices for specific geographic locations. If the modeler is planning to develop a global ionospheric scintillation model, they should always keep in mind that their algorithm should be sensitive to the geographic locations, solar activity as well as local weather. This can be achieved by having different derivation algorithm for each and every geographic location, and the final algorithm should combine all these sub-algorithm to demonstrate the global scintillation response.

2.2 Geometrical error correction and lower elevation multipath

It is better to have the latest data for modeling new events, but, in case of new data unavailability, the previous data may be from the previous solar cycle [6]. The latest data are generally close to the new solar/or geomagnetic event, and they are from the same solar cycle activity period. Therefore, the scintillation models based on the latest data are comparatively closer to the new observations in comparison to the data from the previous solar cycles. The advent of scintillation effect on the trans-ionospheric radio-wave signal lower elevation angles ($\leq 20°$) is being discarded from the data to minimize the effect of multipath. Multipath can occur on any elevation angle. To reduce the high-elevation angle, multipath scintillation receiver's antenna must be setup to minimize CODE and PHASE reflections (multipath), by mounting it away from close reflecting surfaces [6]. The multipath effect also depends on the PRN code rate (please see [7] for details). Moreover the multipath effect can appear at higher elevation angles than 20° (please see [8] for details). Multipath effects occurring at any elevation can be minimized by the geometrical corrections in the data [2]. But, to do it one should be well familiar with the other high-altitude structures (such as mountains or high-altitude building, etc.) near the GPS/GNSS receivers' location and filter the datasets reflected from such structures to reduce the multipath effect. To avoid the high-altitude structures, we have to manual check the data and avoid those elevation angles which show unusual enhancement in the scintillation indices. Following Rino [9], ionospheric irregularities orient themselves according to the local geomagnetic field lines. The phase lock loss highly depends on the angle between the geomagnetic field lines and receiver-transmitter line of sight [10, 11]. For example, at the polar ionosphere, ionospheric irregularities often form rodlike structure and orient themselves along the vertical geomagnetic field lines, whereas at the mid and equatorial latitudes, the local geomagnetic field line is merely horizontal to the Earth's surface due to which sheet- or winglike ionospheric irregularity structures frequently appear. If the irregularity orientations are along the geomagnetic field lines, they form a rodlike structure; if the ionospheric irregularity orientation is across the geomagnetic field lines, it may appear either as a wing- or as a sheetlike structure. We are addressing both cases (rodlike structure and the high-latitude and sheet-/winglike structure at the equatorial and mid-latitude) in this book chapter.

Following Rino [9] and Booker [12], phase and amplitude scintillations can be expressed as follows:

$$< \delta\varphi^2 > = r_e^2 \lambda^2 (Lsec\theta) G C_s \frac{q_0^{-2\vartheta+1} \Gamma\left(\upsilon - \frac{1}{2}\right)}{4\pi \Gamma\left(\upsilon + \frac{1}{2}\right)} F(a,b) \tag{1}$$

$$S_4^2 = r_e^2 \lambda^2 (Lsec\theta) C_s Z^{\upsilon - 1/2} \frac{\Gamma\left(\frac{2.5-\upsilon}{2}\right)}{2\sqrt{\pi}\Gamma((\upsilon+0.5)/2)(\upsilon - 0.5)} g\, F(a,b) \tag{2}$$

where r_e is the classical electron radius; λ is the wavelength of the signal; L is the irregularity slab thickness; θ is the satellite zenith angle; C_s is the turbulence strength parameter, which is a function of the fluctuation in the electron density (ΔN/N) in the irregularity slab along the satellite signal path; G is the geometric

S. No.	Elongation parameters' axial ratio	Ionospheric irregularity form
1	a:1 (i.e., b = 1)	Field aligned rods
2	a:a (i.e., a = b)	Sheets elongated both along the magnetic field and in transverse plane coinciding with local L shell
3	a:b and a > b	Like wings
4	a:b and a < b	This combination is impossible, as ionospheric structures cannot have their spread more in transverse direction of the geomagnetic field

Table 1.
Form of the ionospheric irregularities.

factor and is a function of ionospheric irregularity elongation parameters F(a,b), "a" is the elongation parameter of the irregularities along the filed lines and "b" is used for elongation across the geomagnetic field lines; υ is the three-dimensional spectral index; q_0 is the inner scale constant and $Z \rightarrow \frac{\lambda Z_R Sec\theta}{4\pi}$; $Z_R = ZZs/(Z + Z_s)$ and Z_s are the distance to the source; q is the common geometry and propagation factor, which is also a function of elongation parameters a and b; and $<\delta\varphi^2>$ is the phase variance in the satellite signal after passing through the ionospheric irregularity.

Following the assumption of the theory of wave propagation in random medium, it is safe to assume at the signal frequency of interest that only the phase of the signal wave gets distorted [2] as the signal passes through the ionospheric irregularity of slab thickness L, and GPS scintillation receivers can observe the time series of the phase-modulated signal on the ground. Formulas (1) and (2) can be used to simulate the amplitude and phase scintillation. Study of the power spectrum of the ionospheric data improves the estimation of the local ionospheric irregularity form and their orientation with respect to the local geomagnetic field. Several ionospheric parameters such us spectral index, turbulence strength parameter, and phase fluctuations are essential for correcting the data contaminated through the geometrical errors.

Following Eq. (2) discussed in Rino [9], scintillation index is a function of Fresnel filter factor F(a,b) which is a function of elongation parameters "a" and "b." A detailed explanation of this function is discussed in Rino [9]; as it is beyond the scope of this book chapter therefore we are not discussing it in more details. But, for the reader's convenience, we are providing a summary table (please see **Table 1**), which summarizes the different combinations of "a" and "b" giving rise to different ionospheric irregularity shapes.

3. Scintillation modeling needs and discussions

Scintillation modeling provides a general scenario of the ionospheric scintillations' global morphology and occurrence during different solar activity and space weather conditions. It is always not possible to obtain ionospheric scintillation observation during some space weather events, and during such situations, a realistic ionospheric scintillation model can be used to fill the data gaps. On the other hand, data assimilation techniques also use scintillation model to assimilate scintillation observation into the scintillation model in order to improve the forecast.

There are two aspects that influence the occurrence and strength of the ionospheric scintillation. First is the geometrical effect between the radio-wave-emitting satellite and the receiver which can be at any different location, e.g., at the ground;

on a moving object such as vehicles, ships, or airplanes; onboard at any beacon satellites; or on a spacecraft. The geometrical effect is highly dependent on the elevation angle at which the receiver receives the satellite signal, azimuth angle of the receiver, and orientation of the ionospheric irregularity structures with respect to the local geomagnetic field. The geometrical effects also depend on the angle between the signal ray path and the local geomagnetic field. Another important aspect that influences the scintillation morphology and the rate is the combination of several things such as the geographic location of our region of interest, local season, local time, solar activity condition, during a geomagnetic quite day or a geomagnetically disturbed day, etc. Here first we will discuss in details the geometrical influence on the ionospheric scintillation using the illustrative examples and demonstrate the influence of the geometrical correction on the satellite signal.

3.1 Influence of the geometrical effects over the scintillation estimation and the way to remove these errors

In Eq. (1) the height of the ionospheric irregularity from the Earth's surface Z is a function of sec(θ). The zenith angle (θ) is a function of elevation angle (E) and can be expressed as

$$\theta = 90 - \mathrm{E} \tag{3}$$

Using the relation between satellite and zenith angle, Eq. (2) can be simplified as

$$\mathrm{S}_4 \sim \left[\sec\ (90 - \mathrm{E})\ \sec\ (90 - \mathrm{E})^{\upsilon - 1/2}\right]^{1/2} \tag{4}$$

As we mentioned earlier, υ is a three-dimensional ionospheric irregularity spectral index [9]. One-dimensional spectral index (p) is related to the three-dimensional spectral index as p = 2υ−1. If we simplify Eqs. (3) and (4), we will get to a direct dependence relationship between scintillation indices and zenith angle. Following Priyadarshi and Wernik [13], we can derive the spectral index p by using the log-log relationship of the scintillation index observation and cosecant of the satellite elevation angle [13]:

$$S_4 \sim \csc(E)^{(p+2)/4}\ \mathrm{F}(\mathrm{a}, \mathrm{b}) \tag{5}$$

From Eq. (5) it is clear that the scintillation index is a power-law function of cosecant of the elevation angle with the power one-dimensional ionospheric irregularity spectral index (p) [13]. In order to simplify it more and avoid the dependence on the filter factor F(a,b), which is a complicated function of the ionospheric irregularity elongation parameters and which makes the overall ionospheric irregularity orientation dependence very complicated [13], we have considered the ionospheric radio-wave propagation environment as isotropic [9], and this turns F(a, b) = 1. Now if we plot log-log maps for S_4 and Sin (E) angle, we can calculate the one-dimensional spectra index (p) from this relationship. Once we have spectral index, we can correct the scintillation for the geometry of propagation between the receiver and the transmitter using the Eq. (6):

$$\mathrm{S}_{4\_\mathrm{corrected}} = \mathrm{S}_{4\_\mathrm{observed}} / \csc(\mathrm{E})^{(\mathrm{P}+2)/4} \tag{6}$$

Figure 1 shows the non-corrected ($S_{4\_observed}$) Vs-corrected scintillation indices ($S_{4\_corrected}$) observed from the GPS scintillation receiver GSV 4004b deployed at the Hornsund, Svalbard (76.9718° N, 15.7844° E). The red dots show the amplitude scintillation index, whereas the black dots show the phase scintillation index. If we compare these two figures, the gap between amplitude and phase scintillation index is very less in uncorrected case (on the left). We see that the number of amplitude and phase scintillation observations is reduced in the corrected case (on the right). It is evident from **Figure 1** that after correcting the scintillation data, we come over the geometrical effect for the amplitude as well as phase scintillation. After the geometrical effect correction, we observe two main things. Firstly, there is a reduction in the numerical peak value of the scintillation indices, and secondly, we do see the significant numerical level gap between amplitude and phase scintillation index data after the geometrical correction.

Figure 2 shows the normalized simulated scintillation index map (scintillation index divided by the scintillation index at 25° elevation angle). This map shows the scintillation index simulation for a high-latitude station in Hornsund, Svalbard.

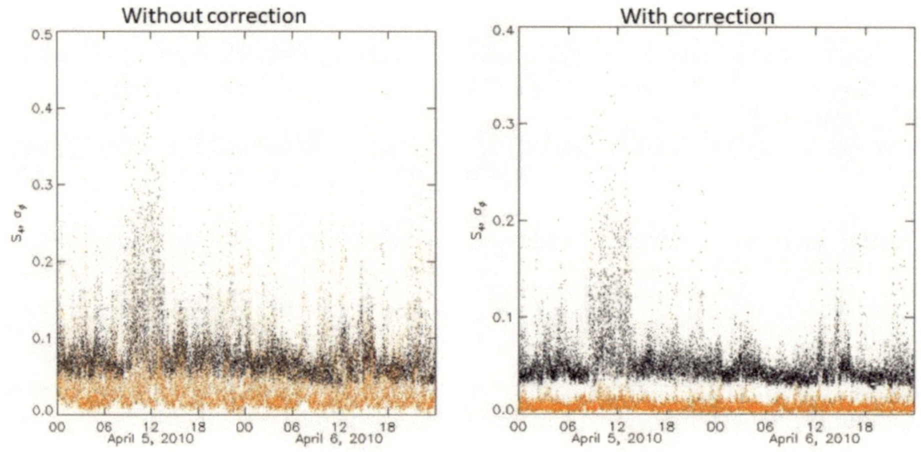

Figure 1.
Amplitude (in orange) and phase scintillation (in black) index observations for Hornsund, Svalbard (76.9718° N, 15.7844° E), (on left) without geometrical effect correction, (on right) with geometrical effect correction.

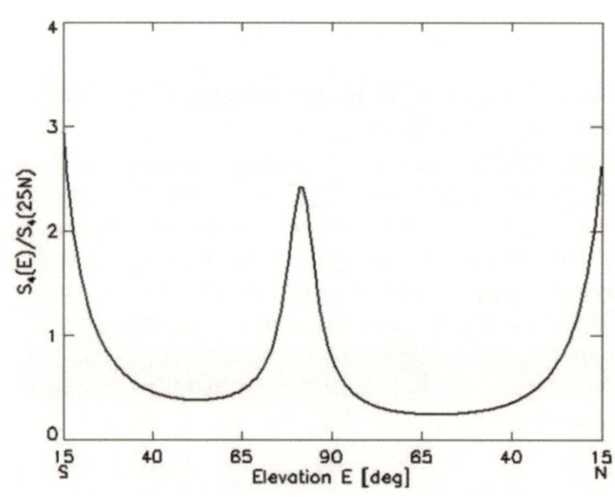

Figure 2.
Normalized simulated amplitude scintillation index for Hornsund, Svalbard, vs elevation angle of the transmitter.

Here we see the scintillation index maximizes between 15 and 20° elevation angles as well as between 70 and 85° elevation angles. At 15–20° elevation angle, the multipath effects are dominant due to which we observe high-value scintillations. On the other hand, between 70 and 85° elevation, the angle orientation of the geomagnetic field line at Hornsund is such that it forms rodlike structure along the geomagnetic field lines. At certain combination of the elevation and azimuth angle, the receiver looks along the geomagnetic field lines. Due to such ionospheric irregularity structure orientation, GPS/GNSS signal travels more distance through the ionospheric irregularity; consequently, we observe the high numerical value of the scintillation indices at this elevation angle range.

To demonstrate the simultaneous impact of the ionospheric irregularity structure and its orientation with respect to the local geomagnetic field lines, we have simulated amplitude scintillation index for a mid-latitude regions Weihai (geographic latitude 37.53°, geographic longitude 122.05°). We have simulated a mid-latitude region for simulating the scintillation index (please see **Figure 3**) using the method discussed in Priyadarshi and Wernik [13] because at the mid-latitude orientation of the geomagnetic field line is neither completely horizontal (alike the equatorial region) nor completely vertical (as at the polar region).

Such slant orientation of the geomagnetic field with respect to the local Earth's surface allows the ionospheric irregularities to evolve and settle in several forms. The orientation parameters along and across the geomagnetic field lines are shown in **Figure 3a** and **b**. The top-left panel shows the rodlike structure, in which we see that for lower elevation angle (≤10°), the scintillation index is very high. As the elevation angle gradually increases, there is a reduction in the scintillation index, and the scintillation index is minimal for higher elevation angle (≥70°). It means that when the spread of the ionospheric irregularity is not significant neither along

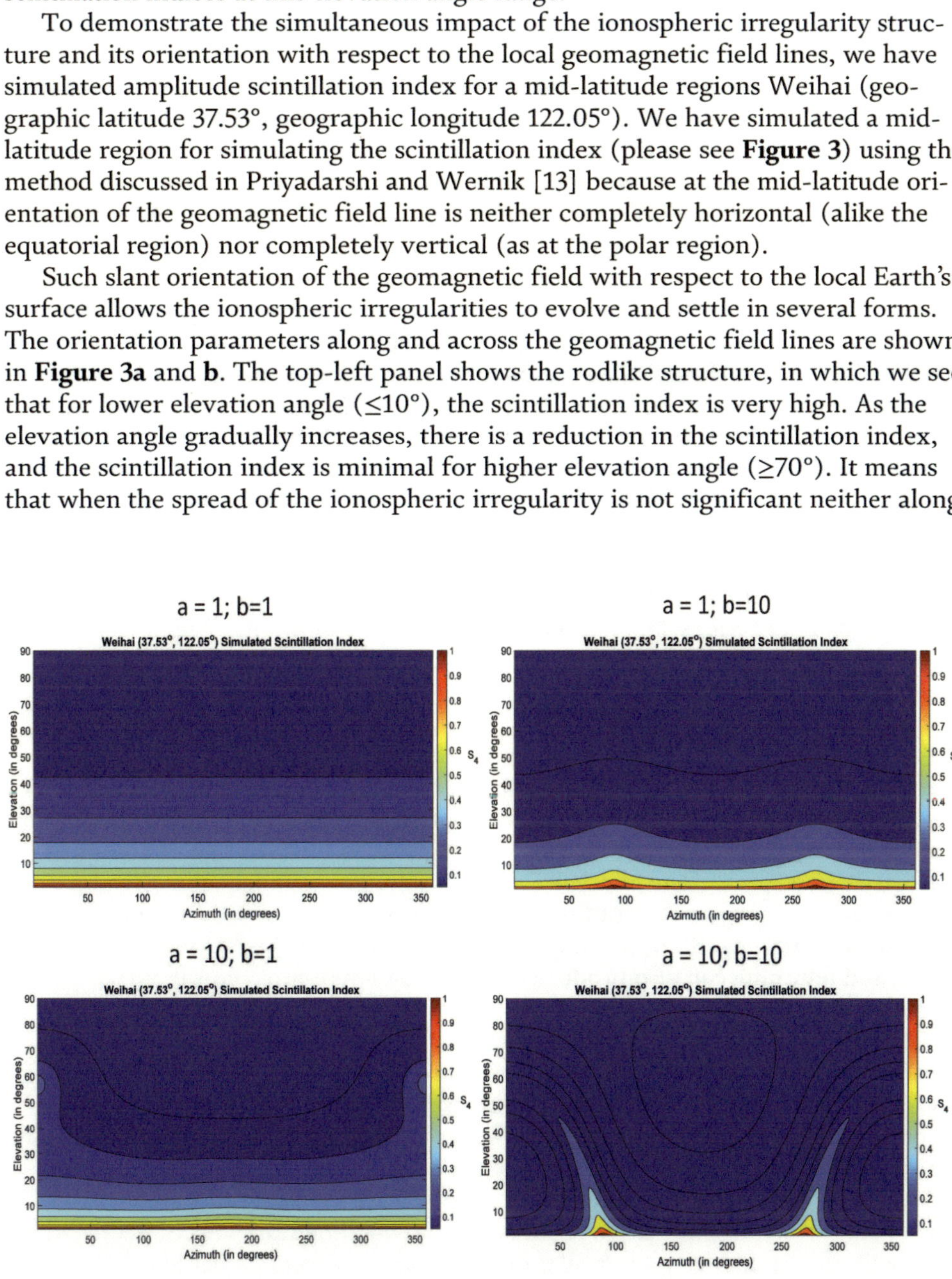

Figure 3.
Simulation scintillation indices for the different combinations of the ionospheric irregularity elongation parameters: rodlike (top-left), winglike (top-right and bottom-left), and sheetlike (bottom-right).

nor across the geomagnetic field lines, then the strength of the scintillation solely depends on the elevation angles. When the spread of the ionospheric structures are more either along (bottom-left) or across (top-right) the geomagnetic field lines, then the numerical value of the scintillation index depends on the combination of the elevation angle and azimuth and on the orientation of the irregularities with respect to the local geomagnetic field lines. As we can see that even for the winglike structures, the scintillation behavior is different for the irregularity spread along and across the geomagnetic field lines. For a = 1 and b = 10 (top-right figure), it shows the decrease in the scintillation with increasing elevation angle, and it also shows scintillation maximizes near two azimuth angles nearly 90° and 270°, respectively. At these two azimuth angles, the line of sight of the observer goes close to the vertical direction of the geomagnetic field lines; ionospheric irregularity structure orientations are vertical and in the direction of the geomagnetic field line. Due to such vertical irregularity orientation, we observe the sharp enhancement in the scintillation index. The third case (a = 10 and b = 1; bottom-left) seems quite similar to the case discussed in the figure at the top-left (a = 1 and b = 1), except 300° > azimuth < 50°. In the fourth case (a = 10 and b = 10; bottom-right), there are two spikes at the azimuth 90° and 270°; these spikes are due to the direction of the geomagnetic field effect as well as a gradual decrease in the scintillation with increasing elevation angle. In summary, a certain combination of azimuth and elevation angle and orientation of the geomagnetic field allow radio-wave satellite signal passing more through the ionospheric irregularity structure.

3.2 Limitations as well as efficacy in modeling the scintillation, due to solar activity, local seasons, local time, and geomagnetic activity

It is very challenging to model ionospheric scintillation during different solar activity, season, local time, and geomagnetic activity conditions. The first limitation for empirical modeling may be due to the specific data unavailability during the particular geomagnetic, solar, and local seasons/time conditions. The modeler should be very careful in making the scintillation model sensitive to the several ionospheric anomalies that appear at different geographic locations and seasons. For example, winter anomaly causes more ionospheric scintillation production during the winter months than the summer months in mid-latitude regions. ±20° of the magnetic equator is the equatorial anomaly region. In the equatorial anomaly region, we observe strong ionospheric scintillation as a lot of scintillation-producing ionospheric irregularity deposit in these regions due to the fountain effect. At the equatorial electrojet regions which are at ±3° of the magnetic equator, we also observe severe ionospheric scintillations. Other events such as X-rays, sudden ionospheric disturbances (SID); protons, polar cap absorption (PCA); and geomagnetic storms and lightning can also produce a significant amount of scintillation-producing ionospheric irregularities.

On the other hand, at the polar latitudes, it is widely believed that the phase scintillation index is more sensitive to the solar/geomagnetic events than the amplitude scintillation index [14–17]. Let us understand this by an illustrative example. Following the South Pole scintillation model developed by Priyadarshi et al. [18], we have produced modeled amplitude and phase scintillation maps for two geomagnetic storms, which occurred in the year 2014 (please see **Figures 4** and **5**). This South Pole empirical scintillation model uses two b-spline functions of degree 4, along with the Ap index and scintillation observation recorded over South Pole GSV4004 scintillation receiver (location, 89.99° geographic latitude; 93.77° geographic longitude; geomagnetic coordinate, 73.5°, 127.8°). Since the model uses a single GPS receiver data, therefore, MLT time lags by 4 hours and 22 minutes to the

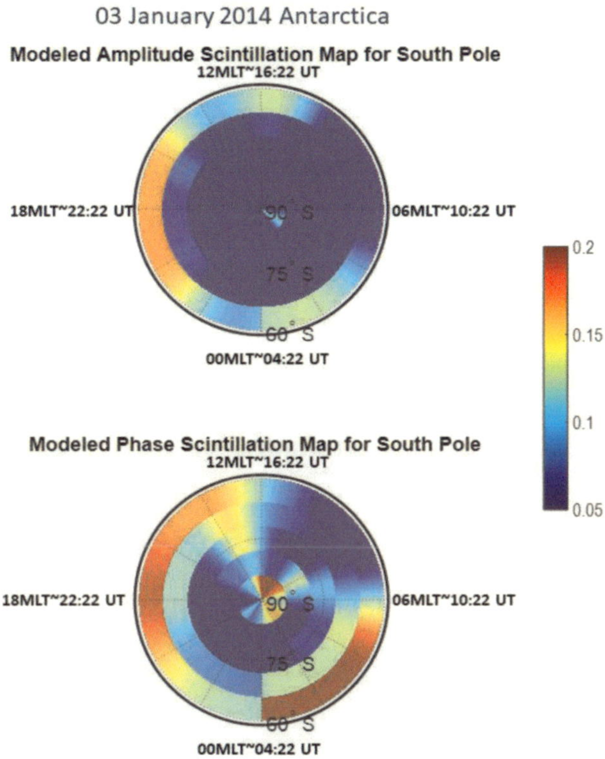

Figure 4.
Modeled amplitude and phase scintillation index over Antarctica during a moderate geomagnetic storm that occurred on 03 January 2014.

universal time hours (UT). As evident from **Figure 4** (between 60 and 70° MLAT), during a weak geomagnetic storm (−30 > =Dst > = −50 nT) that occurred on 03 January 2014, the amplitude scintillation index shows quite similar fluctuations with the phase scintillation index (please see 16–20 MLT at 60–70° S CGMLAT; 0–6 MLT at 60–70° S CGMLAT, and 10–14 MLT at 60–70° S CGMLAT). But, overall phase scintillation values were relatively higher than the amplitude scintillation index. However, during a strong geomagnetic storm (Dst < = − 100 nT) that occurred on 27 February 2014, we see that some part of the amplitude scintillation index shows resemblance with the phase scintillation index map, for instance, 0–2 MLT and 16–18 MLT at 60–70° CGMLAT, but, in the rest of the part near the dusk regions at all the CGMLATs 60–90° S CGMLAT, phase scintillation indices are much higher and high in occurrence than the amplitude scintillation index. It should be noted here that it is not compulsory that S_4 and σ_φ have similar character of variations. In general it depends on the scale of the ionospheric turbulences. In case of kilometers size, the refractive scintillations of the phase are predominant, but amplitude scintillations are minimal. In case of hundred meter-sized turbulence, both refractive and diffractive mechanisms are present, and, hence S_4 and σ_φ indices have high similarity of variations (please see [19] for details).

Space physicist generally explains this that during such strong geomagnetic storms, large-scale ionospheric irregularity structures often enter from the cusp region, and following the convection cells, they pass through the deep inside the polar cap and exit near the magnetic midnight or dusk sector of the polar region. This explanation seems true, but, following the scintillation theory described in Wernik et al. [1], when phase scintillation index keeps on increasing but amplitude

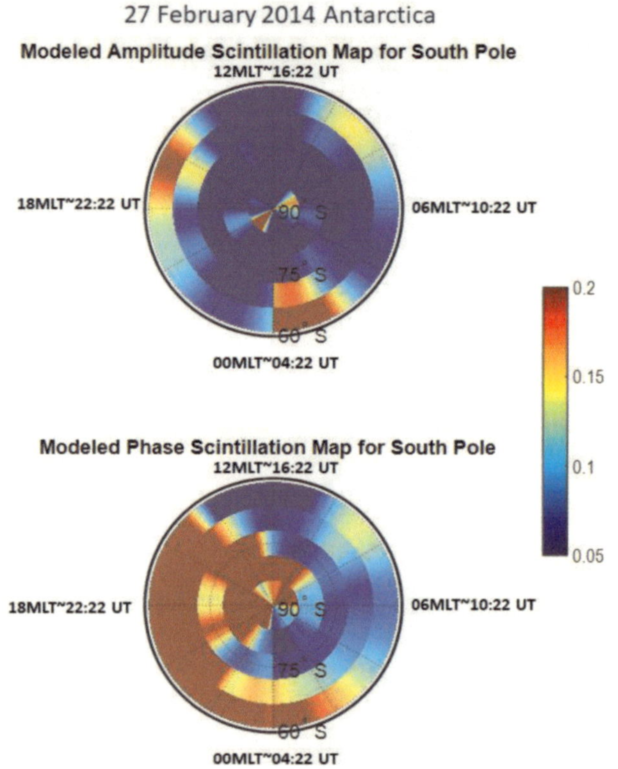

Figure 5.
Modeled amplitude and phase scintillation index over Antarctica during a strong geomagnetic storm that occurred on 27 February 2014.

scintillation seizes to increase, then at this instance, the ionospheric irregularity plasma waves are not more coherent, and these fluctuations cannot be considered as real scintillation due to lack of interference between two noncoherent plasma waves at high-phase fluctuations [1]. For this situation the modeler must optimize the numerical scales of the observed scintillation indices such that the amplitude and phase scintillation indices show similar fluctuation to each other. Priyadarshi et al. [20] used an optimized numerical scale for the amplitude and phase scintillation indices observed during a geomagnetic storm that occurred on 27 February 2014. Ionospheric scintillation index optimization is a way in which by using different amplitude and phase scintillation variation scale it helps us see the same trend variation in the amplitude/phase scintillation indices. In the event discussed in Priyadarshi et al. [20], phase scintillation index variations were optimized between 0.05 and 0.5, and amplitude scintillation index variations were optimized between 0.05 and 0.2. The dayside cusp region amplitude and phase scintillation indices gave similar information at different numerical scales. This study also demonstrated that the amplitude scintillation index is also a useful scintillation index if the proper numerical scale is chosen. Large-scale ionospheric irregularities such as storm-enhanced density (SED) and tong of ionization (TOI) were not found necessarily producing ionospheric scintillation [20].

4. Summary

In general it is more typical when the carrier-phase loss of lock happens due to the sharp signal-to-noise ratio fading (please look [21, 22] for more details). In this book chapter, we have discussed some peculiar modeling tricks and tips of ionospheric scintillation. If taken care well, the model limitations caused due to geometrical effects such as multipath, elevation angle dependence, and form of ionospheric irregularities can be reduced, and the modeled results would provide an error-free estimation of the scintillation indices. Scintillation observations are one of the scintillation model input. By studying the log-log variation of the scintillation index, we can derive the scintillation index correction parameters as discussed in Section 2 [13]. These scintillation correction parameters are used to correct the scintillation model input data. We have demonstrated the geometric corrections between the satellite and receivers (see **Figure 1**) and its importance to use corrected scintillation indices as model input for modeling the ionospheric scintillation indices. Ionospheric scintillation highly depends on the orientation as well as the spread of the ionospheric irregularities with respect to the local geomagnetic field lines. In order to get the real and justified model's output, we must use the input scintillation data in our model for all available, local seasons, local time, and solar/geomagnetic activity duration. At certain elevation angles, scintillation indices show unusual enhancement (please see **Figure 2**); it is because the signal travels long distances through the ionospheric irregularity structures at the lower elevation angle, and at certain elevation angle, the receiver look along the vertical orientation of the ionospheric irregularities with respect to the local geomagnetic field line. The orientation of the ionospheric irregularities structures influences the ionospheric scintillation intensity. We have compared the four types of combination of ionospheric irregularity structures using four different combinations of elongation axial ratios "a" and "b" (please see **Figure 3** and its related discussion). **Figure 3** demonstrated that at a few combinations of azimuth and elevation angles, the orientation of the geomagnetic field allows radio-wave satellite signal passing more through the ionospheric irregularity structure. Apparently, scintillation intensity enhances at these locations. Through two modeled South Pole scintillation maps during the moderate and strong geomagnetic storms (case studies related to **Figures 4** and 5), we have demonstrated that the adoption of the scintillation model provides its high case sensitive to the latitudinal anomalies. For example, winter anomaly which occurs at the mid-latitude ionosphere that causes more scintillation occurrence in winter months as compared to the summer months enhanced scintillation ±20° to either side of the geomagnetic equator and EEJ effect which occurs ±3° on either side of the geomagnetic equator. For high-latitude scintillation modeling, it is essential to consider the ionospheric ionization difference during the sunlit months and dark months. The importance of optimizing the numerical scale of phase and amplitude scintillation indices at the polar latitudes is also discussed. In summary, the presented book chapter discusses all the factors that significantly influence the ionospheric scintillation and possible methods to minimize the estimation errors that caused them. Physics-based models are good to produce the global morphology of the ionospheric scintillation, but, they often fail to produce the exact scintillation index during active solar events. Therefore, empirical models are better as they use the physics-based model as background, but they keep on using the real observations as the model input. These joint efforts work in the derivation and development of a decent ionospheric scintillation model, which can produce equivalent scintillation indices for all the geographic latitudes, local weather, local time, and solar/geomagnetic activity conditions.

Acknowledgements

The scintillation data used in this book chapter has been thankfully obtained from the Polish Polar Station Hornsund, Svalbard. We also thank the Space Research Centre, Polish Academy of Sciences (SRC-PAS) for requesting this data.

Conflict of interest

S. Priyadarshi is the principal as well as the corresponding author of this book chapter. The text, as well as figures presented in this book chapter, has not been published anywhere else before.

Author details

Shishir Priyadarshi
Department of Climatology and Atmospheric Protection, University of Wroclaw, Poland

*Address all correspondence to: shishir.priyadarshi@uwr.edu.pl

IntechOpen

© 2020 The Author(s). Licensee IntechOpen. This chapter is distributed under the terms of the Creative Commons Attribution License (http://creativecommons.org/licenses/by/3.0), which permits unrestricted use, distribution, and reproduction in any medium, provided the original work is properly cited.

References

[1] Wernik AW, Alfonsi L, Materassi M. Ionospheric irregularities, scintillation and its effect on systems. Acta Geologica Polonica. 2004;**52**(2):237-249

[2] Wernik AW, Alfonsi L, Materassi M. Scintillation modeling using in situ data. Radio Science. 2007;**42**:RS1002. DOI: 10.1029/2006RS003512

[3] Singleton DG. Dependence of satellite scintillations on zenith angle and azimuth. Journal of Atmospheric and Terrestrial Physics. 1970;**32**(5): 789-803. DOI: 10.1016/0021-9169 (70) 90029-2

[4] Briggs BH, Parkin IA. On the variation of radio star and satellite scintillations with zenith angle. Journal of Atmospheric and Terrestrial Physics. 1963;**25**(6):339-366. DOI: 10.1016/ 0021-9169(63)90150-8

[5] Forte B, Radicella SM. Comparison of ionospheric scintillation models with experimental data for satellite navigation applications. Annales de Geophysique. 2005;**48**(3):505-514

[6] Current IGS Site Guidelines. Available from: https://kb.igs.org/hc/en-us/articles/202011433-Current-IGS-Site-Guidelines

[7] Padma B, Kai B. Performance analysis of dual-frequency receiver using combinations of GPS L1, L5, and L2 civil signals. Journal of Geodesy. 2019;**93**:437-447. DOI: 10.1007/ s00190-018-1172-9

[8] McCaffrey AM, Jayachandran PT. Spectral characteristics of auroral region scintillation using 100 Hz sampling. GPS Solutions. 2017;**21**:1883-1894. DOI: 10.1007/s10291-017-0664-z

[9] Rino CL. A power law phase screen model for ionospheric scintillation: 1. Weak scatter. Radio Science. 1979; **14**(6):1135-1145. DOI: 10.1029/RS014 i006p01135

[10] Ma G, Maruyama T. A super bubble detected by dense GPS network at east Asian longitudes. Geophysical Research Letters. 2006;**33**:L21103

[11] Demyanov VV, Yasyukevich Yu V, Ishin AB, Astafyeva EI. Effects of ionosphere super-bubble on the GPS positioning performance depending on the orientation relative to geomagnetic field. GPS Solutions. 2012;**16**:181-189. DOI: 10.1007/s10291-011-0217-9

[12] Booker HG, Ratcliffe JA, Shinn DH. Diffraction from an irregular screen with applications to ionospheric problems. Philosophical Transactions of the Royal Society. 1950;**242**(856): 579-607. DOI: 10.1098/rsta.1950.0011

[13] Priyadarshi S, Wernik AW. Variation of the ionospheric scintillation index with elevation angle of the transmitter. 2013:**61**(5):1279-1288. DOI: 10.2478/s11600-013-0123-3

[14] Spogli L, Alfonsi L, De Franceschi G, Romano V, Aquino MHO, Dodson A. Climatology of GPS ionospheric scintillations over high and mid latitude European regions. Annales de Geophysique. 2009;**27**(9):3429-3437

[15] Li GZ, Ning BQ, Ren ZP, Hu LH. Statistics of GPS ionospheric scintillation and irregularities over polar regions at solar minimum. GPS Solutions. 2010;**14**(4):331-341

[16] Prikryl P, Jayachandran PT, Mushini SC, Pokhotelov D, Mac-Dougall JW, Donovan E, et al. GPS TEC, scintillation and cycle slips observed at high latitudes during solar minimum. Annales de Geophysique. 2010;**28**(6): 1307-1316

[17] Moen J, Oksavik K, Alfonsi L, Daabakk Y, Romano V, Spogli L. Space weather challenges of the polar cap ionosphere. Journal of Space Weather and Space Climate. 2013;**3**: A02. DOI: 10.1051/swsc/2013025

[18] Priyadarshi S, Zhang QH, Ma YZ, Wang Y, Xing ZY. Observations and modeling of ionospheric scintillations at South Pole during six X-class solar flares in 2013. Journal of Geophysical Research: Space Physics. 2016;**121**: 5737-5751. DOI: 10.1002/2016JA022833

[19] Bhattacharrya A et al. Nighttime equatorial ionosphere: GPS scintillations and differential carrier phase fluctuations. Radio Science. 2000;**35**(1): 209-224

[20] Priyadarshi S, Zhang QH, Ma YZ. Antarctica SED/TOI associated ionospheric scintillation during 27 February 2014 geomagnetic storm. Astrophysics and Space Science. 2018; **363**:262. DOI: 10.1007/s10509-018-3484-x

[21] Tiwari R, Strangeways HJ. Regionally based alarm index to mitigate ionospheric scintillation effects for GNSS receivers. Space Weather. 2015;**13**:72-85. DOI: 10.1002/2014SW001115

[22] Kinter PM, Ledvina BM, de Paula ER. GPS and ionosphere scintillations. Space Weather. 2007; **5**:S09003. DOI: 10.1029/2006SW000260

Section 3

Satellites Missions and Technologies

Chapter 7

Earth Observation Technologies: Low-End-Market Disruptive Innovation

Silvia Rodriguez-Donaire, Miquel Sureda, Daniel Garcia-Almiñana, Eloi Sierra, Jose S. Perez, Peter C.E. Roberts, Jonathan Becedas, Georg H. Herdrich, Dhiren Kataria, Ronald Outlaw, Leonardo Ghizoni, Rachel Villain, Alexis Conte, Badia Belkouchi, Kate Smith, Steve Edmondson, Sarah Haigh, Nicholas H. Crisp, Vitor T.A. Oiko, Rachel E. Lyons, Stephen D. Worral, Sabrina Livadiotti, Claire Huyton, Luciana A. Sinpetru, Rosa M. Domínguez, David González, Francesco Romano, Yung-An Chan, Adam Boxberger, Stefanos Fasoulas, Constantin Traub, Victor Jungnell, Kristian Bay, Jonas Morsbøl, Ameli Schwalber and Barbara Heißerer

Abstract

After decades of traditional space businesses, the space paradigm is changing. New approaches to more efficient missions in terms of costs, design, and manufacturing processes are fostered. For instance, placing big constellations of micro- and nano-satellites in Low Earth Orbit and Very Low Earth Orbit (LEO and VLEO) enables the space community to obtain a huge amount of data in near real-time with an unprecedented temporal resolution. Beyond technology innovations, other drivers promote innovation in the space sector like the increasing demand for Earth Observation (EO) data by the commercial sector. Perez et al. stated that the EO industry is the second market in terms of operative satellites (661 units), micro- and nano-satellites being the higher share of them (61%). Technological and market drivers encourage the emergence of new start-ups in the space environment like Skybox, OneWeb, Telesat, Planet, and OpenCosmos, among others, with novel business models that change the accessibility, affordability, ownership, and commercialization of space products and services. This chapter shows some results of the H2020 DISCOVERER (DISruptive teChnOlogies for VERy low Earth oRbit platforms) Project and focuses on understanding how micro- and nano-satellites have been disrupting the EO market in front of traditional platforms.

IntechOpen

Keywords: disruptive innovation, low-end market, micro- and nano-satellites, new space, Earth Observation

1. Introduction

Although Earth Observation (EO) started as an activity exclusively affordable for governments or big players in space with vast financial resources to sustain expensive programmes, it is no longer an exclusive and expensive industry. It allows the emergence of start-ups and spin-offs from academia and emerging countries that are the foundations of the *New Space*. This phenomenon, known as the democratization of space, changes the accessibility, affordability, and commercialization of space products and services to companies of all types and sizes [1].

According to [2], New Space can be understood as a disruptive trend whose aim is to transform space into a commodity by taking advantage from the joint between Information Technology (IT) and EO. Even though its origins were in Silicon Valley, the trend is now extended worldwide.

Regarding the Union of Concerned Scientists (UCS) satellite database, 1980 operational satellites were orbiting the Earth at the end of April 2018, with 684 of these aimed to EO [3]. This represents a growth of 250% compared to January 2014, when there were only 192 active EO satellites. From this huge increase, it is clear that EO data acquisition is an emerging market. With the number of companies growing year-by-year and optimistic forecasts, it can be reinforced that "EO is on its earlier days and there are still a lot of improvements to do and problems to solve" [4].

Looking at the new EO-based markets, it is observed in [2, 5] some signs of potential disruptive innovation in the space sector. Some technological drivers promote this innovation. For instance, low cost access to earth imagery; availability of high-quality spatial, spectral and temporal imagery; innovations in

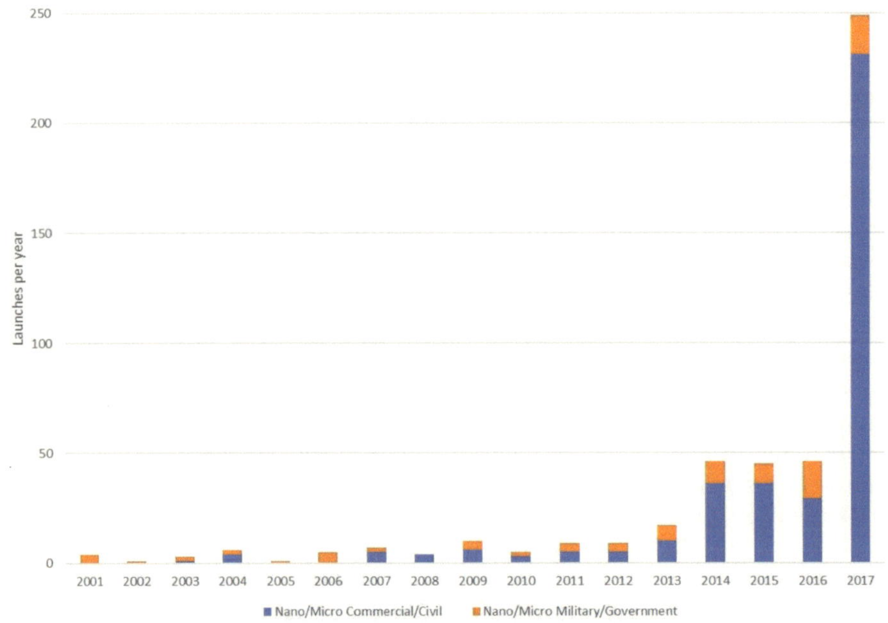

Figure 1.
Micro- and nano-satellites launched between 1997 and 2017, classified by sectors (own elaboration).

computer science, like cloud computing and machine learning; and some specific programmes, like Copernicus, that transfer high technology development from governmental programmes to other industries and services.

Beyond technological drivers, other drivers promote the disruptive innovation in the space sector, such as the sharing economy, the increasing demand of the commercial sector—like smart cities—and the government interests in environmental monitoring. For example, Spaceflight offers companies a global launching opportunity, working with almost every launch vehicle provider on the planet.

All of these drivers encourage new companies—like Skybox, SpaceX, OneWeb, Telesat, Planet, and OpenCosmos, among others—to develop new business models that make space more accessible and affordable for nongovernmental organizations on shorter periods.

Figure 1 summarizes the total number of micro- and nano-satellites launched per year between 2001 and 2017. Micro-satellites are loosely defined as any satellite weighting between 10 and 100 kg, while nano-satellites weight less than 10 kg. They have been classified by sectors to emphasize the huge increase of commercial micro- and nano-satellites launched in recent years compared to those launched by defense departments and governments.

2. Hypotheses: disruption in EO technologies

The term *disruptive innovation* was popularized in 2003 by Clayton M. Christensen, professor at Harvard Business School. In [6], he distinguished between sustaining and disruptive technologies and later, in [7], it replaced the term *technology* with *innovation*, since disruption does not come from technology but from businesses. According to [7], sustaining innovations foster improved product performance, while disruptive innovations bring to the market a very different value proposition, with a performance that is initially below the mainstream products, but with low prices or unique features that compensate for it.

Additionally, in [6], a distinction between low-end-market and new-market innovations is made. Low-end-market innovations are those that do not result in

Factors affecting innovation	Space situation
Challenging objectives and attractive environments (+)	Space missions remain technically very challenging and their components and technologies are still one-off prototypes, custom-designed, and optimized for specific missions
Closed sector (−)	Space is a closed sector with little exchange of resources outside of aerospace and defense. However, innovations, especially the disruptive ones, appear from the intersection of domains and disciplines
Risk adversity (−)	Space activities are high-risk efforts and they do not offer opportunities for error corrections after launch. This leaves little freedom for innovation and leads to a risk-averse culture
Highly skilled workforce (+)	Space workforce is highly educated and mobile and has a diverse cultural background
High entrance barriers and open competitive markets (−)	Without open competitive markets, space innovations are likely considered useless for businesses. Additionally, high entry barriers and huge launching costs reduce the stimuli of industrial and private sectors to invest in space innovations

Table 1.
Factors affecting space innovation [5].

better product performance but offer lower prices, such as Walmart and its cheap retailing malls. On the other hand, new-market innovations, like the iPod, serve new users who had not owned or used the previous generation of products.

Christensen approached disruptive innovations from the point of view of both management and industry. However, his recommendations are kept at industrial level. Despite his product performance and business strategy analysis, his definition does not identify the innovation characteristics, since they are intrinsic rather than external factors that change over time, like customer perception or government regulations.

In [5], the concept of innovation is applied to the space environment. The author stated that some factors would affect the likelihood of innovation within the European space sector. **Table 1** summarizes these factors, dividing them between those that promote space innovation (+) and those that prevent it (−).

In [8], the previous concept of disruptive innovation is refined by identifying three innovation characteristics: functionality, discontinuous technical standards, and ownership models. His definition broadens the meaning of low-end market and new market innovations.

Low-end market innovations *are those with discontinuous technical standards that disrupt markets by using new, less costly materials or new production processes in the creation of existing technologies* [9, 10] *or new forms of ownership. These forms dictate how innovations are received in a marketplace, as they establish prices and innovation-related services among others* [11].

New-market innovations *are those with a disruptive functionality that provides the user with the ability to undertake a new behavior or accomplish a new task that was impossible before* [12–14].

Taking the above signs of innovation in the space sector and following the strategy developed in [7, 8], recent micro- and nano-satellite EO missions seem to show the key characteristics of disruptive innovation. **Table 2** summarizes the characteristics of micro- and nano-satellites as disruptive innovations according to different authors.

By combining the above-presented characteristics of disruptive micro- and nano-satellite innovations with the main specificities of the EO space market, a set of six hypotheses for micro- and nano-space market disruption has been developed

Characteristics of disruptive innovations	Christensen and Raynor [7]	Summerer [5]	Nagy et al. [8]	Denis et al. [2]
High level of risk		■		■
Discontinuous technical standards (simplicity)		■	■	■
Accessibility	■	■		■
Enabling new market opportunities	■	■		■
Inferior performance	■			
Performance improvement		■		■
Disruptive functionality			■	
Affordability	■	■		■
Forms of ownership		■	■	■

Table 2.
Characteristics of micro- and nano-satellites as disruptive innovation in space [2, 5, 7, 8].

Characteristics of disruptive innovations	Hypothesis to test	Hypothesis label
Standardization	The space sector has a low level of risk acceptance, which leaves little freedom for innovation. However, micro- and nano-satellites provide simplicity and standardization in terms of design and manufacturing, This leads to a higher level of risk acceptance and, consequently, more innovation	1
New market opportunities	Data accessibility and technology standardization are essential conditions to open new market opportunities	2
Performance	Micro- and nano-satellites improve their performance in a pace that meets market needs even though they have an inferior performance than those of traditional EO spacecraft	3
Affordability	Traditional, established space companies are ignoring the market due to very low-profit margins. This fact leaves room for new entrants with totally different business models. These new actors bet on low-cost technology to produce more affordable space systems for Earth Observation	4
Ownership forms	Recent evolutions in micro- and nano-satellite technologies are affecting the forms of ownership and operability of EO systems, which were formerly owned by governments or public organizations	5
Disruptive functionality	Micro- and nano-satellite missions offer disruptive functionalities that provide novel products or services that were unthinkable or impossible with traditional spacecraft missions	6

Table 3.
Summary of studied hypothesis related to the disruptive innovation characteristics.

and presented in **Table 3**. In this section, all mentioned characteristics are tested to verify if the authors' hypotheses are true in order to clarify whether micro- and nano-satellites are disruptive for the EO market.

3. Analysis: disruption in EO technologies

In this section, the analysis of the six hypotheses stated in **Table 3** for micro- and nano-space market disruption has been done. The first hypothesis is related to space market standardization, the second hypothesis is related to market opportunities, the third hypothesis is related to micro- and nano-satellite performance, the fourth hypothesis is related to the affordability of the new space technologies for EO, the fifth hypothesis is related to the forms of ownership and operability of EO systems, and finally, the sixth hypothesis is related to disruptive functionalities that provide novel products.

3.1 Hypothesis 1: micro- and nano-satellite simplicity and standardization

H1: *The space sector has a low level of risk acceptance, which leaves little freedom for innovation. However, micro- and nano-satellites provide simplicity and standardization in terms of design and manufacturing. This leads to a higher level of risk acceptance and, consequently, more innovation.*

In mainstream space platforms, each of design, development, and test campaign tends to be almost unique, custom-made for the specific mission. Long project durations and high costs are consequences of the complexity that implies the need of guaranteeing the maximum quality, hence the minimum risk for the mission.

On the other hand, micro- and nano-satellite constellations are based on the concept of standardization, which opens up the possibility of using commercial electronic components and the choice of numerous technology suppliers. In that way, it is possible to create less expensive satellites in shorter periods. Depending on the specifications, a micro-satellite can be built and placed in orbit for a few million euros and a nano-satellite for almost a quarter million. In comparison, the cost of a large satellite can rise to 500 million euros [4, 15].

Apart from the cost and size, the main benefit of micro- and nano-satellites is the time required to design and implement each model. As an average, a micro- or a nano-satellite can be designed, manufactured, and launched within less than 2 years [4, 15]. This means that large constellations of small satellites can be regularly renewed with state-of-the-art systems, ensuring optimal performance even if some units are lost or fail. This is not the case of conventional satellites, which are developed and launched within expensive and long projects that last between 5 and 10 years and, accordingly, cannot afford any failure in the platform without risking the entire mission.

Particularly worthy of mention is the recent emergence of many dedicated micro-launchers designed to place small satellites in orbit. So far, micro- and nano-satellites are launched at marginal costs as "piggyback" payload alongside traditional spacecraft. However, new micro-launcher concepts may be responsible for providing simplicity and standardization to the whole process, lowering launch costs if they demonstrate reliability and good performance [1].

For the stated reasons, H1 can be supported, since micro- and nano-satellite design and manufacturing is focused on simple and standard equipment that eventually may increase the linked risk acceptance.

3.2 Hypothesis 2: new market opportunities

H2: *Data accessibility and technology standardization are essential conditions to open new market opportunities.*

EO is a promising, fast-growing field boosted by a wide range of applications across various economic sectors, including precision farming, natural resource monitoring, oil and gas exploration, meteorology, civil protection, insurance, and urban monitoring [1]. The emergence of low-cost micro- and nano-satellites enabled EO start-ups to attract new markets interested in their tremendous amount of accessible and affordable high-resolution images. Additionally, more and more countries invest in their EO capacity, confirming the soft power dimension of space but also opening new market opportunities for international or regional cooperation [1].

Not only space is becoming more accessible through new launch technology, but also data from programs like US Landsat and Europe's Sentinel program are already available to all. This allows third parties to develop new services and applications over high-quality databases supported by different funding programs. For instance, OneAtlas updated the base map of the whole world with high-resolution imagery without taking any picture or OneWeb plan to use small spacecraft technology to make satellite Internet available on a global scale.

It is clear that some of these new markets are recently gaining access to EO data because it is cheaper than before. However, a very important entry barrier was also the traditional space companies themselves, because data owned and controlled by defense and public organizations were not available at any price.

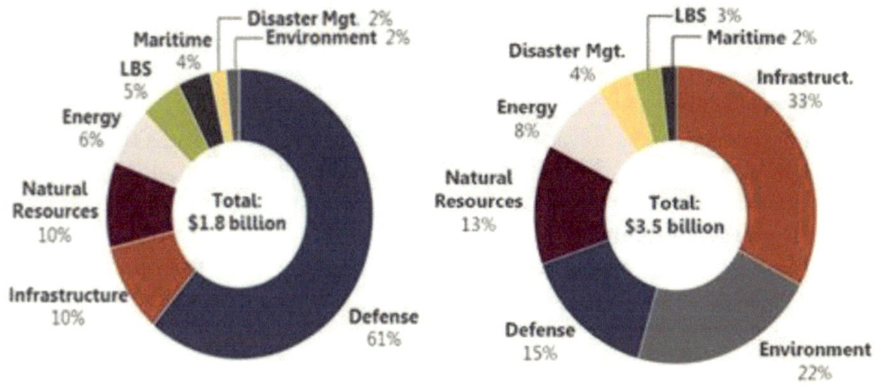

Figure 2.
Commercial EO data market in 2017 (left) and value-added services market in 2017 (right) [1].

In **Figure 2**, it can be seen that in the year 2017, defense represented more than 60% of the commercial data market ($1.8 billion), with infrastructure and natural resources verticals accounting a similar share to each other. These three vertical markets represented 80% of the commercial data market in 2017. Looking to the future, Euroconsult forecasts that the market for commercial EO data is expected to reach $3 billion (5% of the Compound Annual Growth Rate (CAGR)) in 2026 [1].

In the short term, growth is expected to continue to be driven by the defense, with ongoing regional unrest and growing Image Intelligence needs of countries without proprietary military systems. By 2026, the defense is expected to represent 46% of the total market value ($1.7 billion). Therefore, although defense will continue to be the major client for EO imagery, their share will reduce in the coming years. Other applications, such as maritime, infrastructure, and resource monitoring will support growth in the long term. Together with defense, these applications should have a 5% CAGR through 2026. Emerging applications in these sectors such as critical infrastructure monitoring and precision agriculture benefit from more capable satellite systems (i.e., a combination of higher ground resolution with higher temporal resolution). Location-Based Services (LBS) applications, including financial and insurance services, have been slow to develop, but the longer-term outlook for these services remains positive with the availability of new satellite capacity. For LBS applications, greater emphasis is expected to be put on integrated product offerings, emphasizing requiring the development of change detection analytics. In terms of revenue generation by data type, VHR optical is expected to remain the most significant in terms of data sales. More moderate-resolution datasets will be challenged by the availability of free solutions and low-cost systems offering comparable data.

According to [1], in 2016, the market for Value-Added Services (VAS) was $3.5 billion. This discounts the purchase of commercial data to develop geospatial solutions. Key markets for VAS do not mirror those for commercial data sales. Defense, while representing 61% of the commercial data market, only represents 15% of the VAS market; conversely, infrastructure and engineering (which incorporated cartography, cadastre, etc.) is only 10% of the commercial data market but 33% of the value-added market.

According to [1], the reasoning for this is relatively straightforward: defense end-users purchase data with much value-added analytics performed in-house. On the other hand, lower-cost, coarser resolution, and lower geolocation accuracy data can be leveraged with value-adding to form greater value products and services. Environment-monitoring users, for instance, procure limited commercial data but are developing solutions using scientific and coarse resolution data, for example, pollution/aerosol monitoring and climate modeling. Many infrastructure applications for mapping also can be developed by using Landsat and Sentinel data that are free of charge.

In [1], it is also forecast that data also add to the belief that by making coarser-resolution data free, the value-added services industry can leverage this to build greater value services with the potential for two very different businesses: a "high-end" data market to support defense and free/low-cost data sources to support commercial and civil government applications.

For these reasons, H2 would also be supported, since new market opportunities are growing and standardization has been proven as H1.

3.3 Hypothesis 3: micro- and nano-satellites performance

> H3: *Micro- and nano-satellites improve their performance in a pace that meets EO market needs despite having an inferior performance than those of traditional EO spacecraft.*

EO optical imaging satellite performance is defined in terms of spatial and temporal resolution. Spatial resolution relates to the level of detail obtained from an image and can be measured by the Ground Sample Distance (GSD), which is the distance between adjacent pixel centers measured on the ground.

Figure 3 shows the evolution of EO micro- and nano-satellite GSD in the last 20 years. The solid lines depict how the concepts of Medium Resolution (MR), High Resolution (HR), and Very High Resolution (VHR) evolved through time. While MR has maintained constant around 15 m, HR and VHR have decreased to 2 and 0.3 m, respectively.

Dots in **Figure 3** represent GSD values for the EO micro- and nano-satellites analyzed in this research (see Appendix A for details on the data analysis methodology). Cross marks prove that between 1999 and 2013 governmental and defense were almost the only micro- and nano-satellites devoted to obtaining HR and VHR images of the earth. However, 2013 marks a turning point in the EO market, with

Figure 3.
Evolution of EO satellite GSD during the period 1999–2018 (own elaboration).

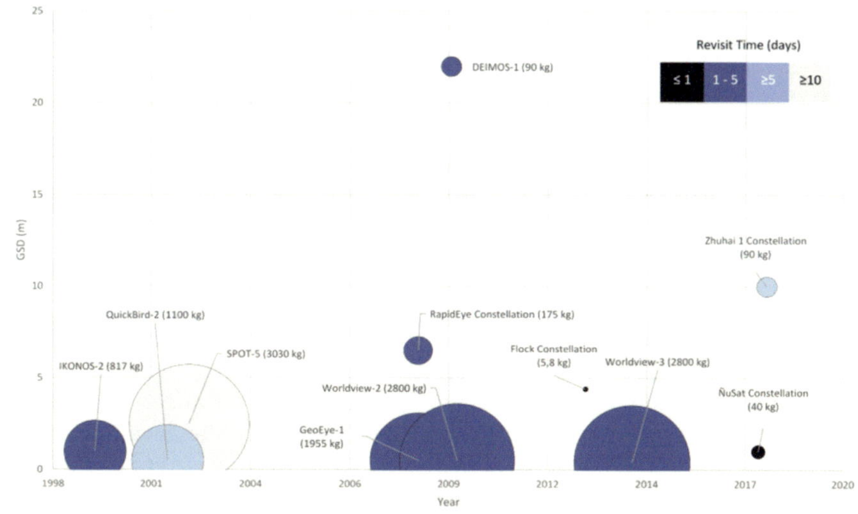

Figure 4.
GSD, year of launching, mass, and revisit time of EO micro- and nano-satellites (own elaboration).

the irruption of private start-ups launching small space platforms able to achieve resolutions from 4 to 1 m [15].

Nano-satellites can obtain GSD below 3 m, thanks to new sensor technologies and the use of Low Earth Orbit and Very Low Earth Orbit (LEO and VLEO). In the range of micro-satellites, GSD around 1 m can be achieved, with perspectives of values even lower in the next years. These values are small enough to have a great interest for many commercial applications, mainly in the agriculture, transportation, energy, and infrastructure markets [1].

If the temporal resolution is also taken into account, micro- and nano-satellites show their huge potential for EO commercial applications. Satellite's revisit time is the time elapsed between observations of the same point on Earth's surface. **Figure 4** summarizes GSD, year of launching, mass, and revisit time of EO micro- and nano-satellites. The amount of satellites displayed in **Figure 4** is smaller than in **Figures 1** and **3** because the information about revisit time was not available for many satellites (see Appendix A for details on the data analysis methodology). Besides, revisit time of Flock (Planet) and ÑuSat (Satellogic) constellations is calculated using their final future configuration.

In **Figure 4**, each circle represents a satellite or a constellation of identical satellites. GSD and launching year can be measured in both axes, while the circle gives information about the mass and revisit time (the bigger, the more massive and the darker, the shortest revisit time). Looking at the characteristics of different satellites, it is easy to see that Flock and the ÑuSat constellations are the only platforms able to provide revisit times lower than one day. This capability makes their data more appealing than any of the other platforms, even having a slightly less spatial resolution. The key for this performance is the possibility to design, launch, and operate constellations of more than 100 satellites, something which seems only possible, thanks to the reduced costs associated to micro- and nano-satellite technology considering the several hundred million dollar cost of traditional EO satellites. These massive micro- and nano-satellite constellations are aiming to transform EO imagery into a commercial product (e.g., analytical solutions from the big data obtained from the constellations), taking benefit of their almost high resolution and their high revisit time.

For all these arguments, H3 would be supported, as performance is increasingly high in new satellites and constellations, whilst it is still far from conventional satellites.

3.4 Hypothesis 4: affordability of the new space technologies for EO

H4: *Traditional, established space companies are ignoring the market due to very low-profit margins. This fact leaves room for new entrants with totally different business models. These new actors bet on low-cost technology to produce more affordable space systems for Earth Observation.*

During the last decades of the twentieth century, EO systems were mainly dedicated platforms owned and operated by public organizations or governments, often at a national level. This status quo was sustained by economic and policy barriers to space commerce. Traditionally, costs associated with satellite development and operation have been extremely high, both at LEO and Geostationary Orbits (GSO). However, platform standardization, continued progress in technology miniaturization, and Components Off-The-Shelf (COTS) are not only leading to cheaper satellite development and launch but also reducing manufacturing time. The possibility of using many small satellites in a constellation is enabling near real-time Earth Observation and addressing the issue of temporal resolution. Consequently, increasingly large amounts of data are being gathered every day.

This novel combination of price reduction and data generation has created in the last decade new business opportunities favoring the emergence of new space companies dedicated to the EO market. These companies base their innovative business models on the generation of near real-time high-resolution images (close to 1 m [15]) that are sold in user-oriented data access platforms (around $1/km^2 [15]).

It is important to note that this new model is mainly ruled by start-ups with substantial investment capacity. **Figure 5** depicts the evolution of investment and the number of start-ups founded in the EO market between 2013 and 2017. It can be seen that almost 60 EO companies were funded in these 5 years. More significantly, the solid line clearly shows an increase in the investment, from less than 5 million

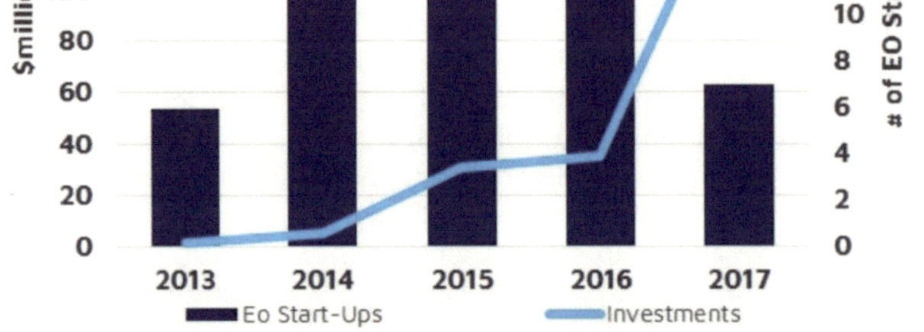

Figure 5.
Earth Observation start-ups and investments during the period 2013–2017 (source [15] and own elaboration).

dollars in 2013 and 2014 to almost 160 million dollars in 2017. According to [15], one of the reasons why new start-ups have been so successful in raising capital may well be because it challenges traditional space enterprises at technology implementation, deployment of spacecraft, innovation in the business model, etc. Some of these new start-ups that embrace open-innovation and knowledge sharing are Planet, Satellogic, PLD Space, Deimos, and GomSpace, among others [2, 5].

For all these reasons, H4 would be supported, since new dealers in the space sector are mainly new companies rather than stablished ones.

3.5 Hypothesis 5: forms of ownership and operability of EO systems

H5: *Recent evolutions in micro- and nano-satellite technologies are affecting the forms of ownership and operability of EO systems, which were formerly owned by governments or public organizations.*

These new opportunities foster innovation and commercial growth, but they also leave room to establish a legal framework. That regulation would aim to maintain a safe and predictable space environment that allows us to face correctly the rapid changes of technology without interrupting the innovation freedom of the space sector.

The disruption in the space market extends the technological improvements. For instance, an increment in the supply of EO imagery would have implications for new business models, lower costs and more flexible ownership models for commercializing imagery.

Emerging start-ups and spin-offs in the space sector are transforming the operability of EO systems owned by governments or public organizations. This transformation extends from the satellites themselves to the data processing and finally the data analysis that represents VAS to commercial and public organizations.

The idea of a "sharing economy" implies a revolution in the ownership of space imagery. All users have access to relevant and free data under a distributed ownership scheme. This trend is being driven by multiple technological innovations, for example, reusable launchers (e.g., SpaceX), online platforms where users can combine different data (e.g., Blacksky), or launcher service platform (e.g., Spaceflight). Nevertheless, there are questions over the sustainability of this ownership model, especially for commercial organizations that need to generate profit. Additionally, there are certain applications related to security or defense where such a shared-ownership model may not be appropriate.

Radiant Earth Foundation is trying to address some of these challenges, such as building a place where the development community can go for earth imagery and geospatial data and with access to market analytics, best practices guides, return of investment methodologies, and discussion of policy issues.

For all these reasons, H5 would also be supported, since new dealers from H4 are also defining new forms of ownership and service.

3.6 Hypothesis 6: disruptive functionality

H6: *Micro- and nano-satellite missions offer disruptive functionalities that provide novel products or services that were unthinkable or impossible with traditional spacecraft missions.*

Although nano- and micro-satellites are dramatically changing the EO market, it cannot be said that they are providing novel products or services to the final data users. Satellite imaging has been used since the early 1970s when the Landsat program started. As stated before, the irruption of new start-ups with novel business models is not based on the generation of new data, but on the accessibility to it and complementing traditional space business models.

Therefore, considering that hypothesis 2 and 4 proved to be true, micro- and nano-satellites in the EO market can be categorized as low-end-market disruptive innovations. This hypothesis is also supported in [15], in which it is stated that new space business models do not drastically change satellite EO business, since both fulfill a similar kind of customer needs. What New Space businesses provide against traditional ones are accessibility, affordability, and commercialization of space products or services to commercial and noncommercial companies.

H6 is therefore not fulfilled.

4. Conclusions

New Space has usually been considered as a disruptive market. Some technological drivers like low-cost, high-quality image, among others, promote innovation, and they encourage new companies to develop new business models that make space more accessible and affordable for nongovernmental organizations on shorter development periods.

This chapter is measuring how disruptive the micro- and nano-satellite innovations are within the EO space market, under a series of hypotheses based on established standards for disruptive innovation [2, 5, 7, 8].

As a result, we have observed that micro- and nano-satellite technologies represent a low-end-market disruptive innovation, since they standardize the production process that reduces the cost of design and manufacturing phases. Additionally, thanks to this standardization, the forms of ownership and operability of EO platforms have changed to a private model. This allows the establishment of lower prices and creation of innovative services that not only open new market opportunities for new business models and data accessibility by commercial companies, but also improve the space market performance even though they have inferior characteristics than those provided by traditional EO spacecraft. As a consequence of all this, it cannot be said that micro- and nano-satellites drastically change satellite EO business, but they provide an accessible and affordable data to commercial and noncommercial companies against traditional ones.

Acknowledgements

The DISCOVERER project has received funding from the European Union's Horizon 2020 research and innovation program under grant agreement No 737183.

Appendix: quantitative data analysis

Most of the data analyzed in this paper were obtained from the Union of Concerned Scientists (UCS) database [1]. This Web includes a listing of nearly 2000 operational satellites orbiting around Earth. Based on this information, only satellites with Earth Observation purposes were selected, which lead to a final number of 394 EO satellites usable for our research. From all these, **Figures 3** and **4** only picture information about those with spatial and temporal resolution values available. The list of used data can be found in **Table 4**.

Name	User	Launch date	Launch mass (kg)	GSD (m)	Revisit time (days)
DLR Tubsat	Government	26/05/1999	45	6	—
IKONOS-2	Commercial	24/09/1999	817	1	2.25
Terra	Government	18/12/1999	4864	15	—
QuickBird-2	Commercial	01/10/2001	1100	0.5	7
Bird 2	Government	22/10/2001	92	25	—
SPOT-5	Government	04/05/2002	3030	2.5	26
BeijinGalaxy-1	Civil	27/10/2005	50	39	—
EROS B1	Military/commercial	25/04/2006	350	0.7	—
RapidEye Constellation	Commercial	29/08/2008*	175	6.5	3.25
GeoEye-1	Commercial	01/09/2008	1955	0.5	3
HJ-1	Government	05/09/2008	470	30	—
Saudisat-2	Government	29/07/2009	35	15	—
DEIMOS-1	Government	29/07/2009	90	22	2
Worldview-2	Commercial	01/10/2009	2800	0.5	2.4
Alsat	Government	12/07/2010*	116	2.5	—
BKA 2	Government	22/07/2012	473	2.1	
Landsat-8	Government	11/02/2013	2623	15	—
DubaiSat-2	Government	21/11/2013	300	1	—
Zhuhai 1 Constellation	Commercial	26/04/2014*	90	10	5
Rising-2	Civil	24/05/2014	41	5	—
Flock Constellation	Commercial	19/06/2014*	5	4	1
Aurora	Commercial	19/06/2014	25	15	—
Worldview-3	Commercial	01/08/2014	2800	0.4	2.75
ASNARO 1	Government	06/11/2014	500	0.5	—
CBNT-1	Commercial	10/07/2015	91	1,5	—
DMC 3	Commercial	10/07/2015*	447	1	—
LAPAN A2	Government	24/09/2015	68	5	—
Bison Sat	Civil	08/10/2015	1	43	—
Athenoxat-1	Commercial	16/12/2015	6	25	—
BIROS	Government	22/06/2016	110	42.4	—
BlackSky Pathfinder	Commercial	26/09/2016	44	1	—
Worldview-4	Commercial	01/11/2016	2485	0.3	4.5
CE-SAT-1	Commercial	22/06/2017	50	1	—
Skysat	Commercial	31/10/2017*	110	0.9	—

**Constellation's launch date stated in the table corresponds to the first launching.*

Table 4.
*List of satellites used in **Figures 3** and **4** [3, 16–19].*

Author details

Silvia Rodriguez-Donaire[1], Miquel Sureda[1], Daniel Garcia-Almiñana[1*], Eloi Sierra[1], Jose S. Perez[2], Peter C.E. Roberts[3], Jonathan Becedas[4], Georg H. Herdrich[5], Dhiren Kataria[6], Ronald Outlaw[7], Leonardo Ghizoni[8], Rachel Villain[2], Alexis Conte[2], Badia Belkouchi[2], Kate Smith[3], Steve Edmondson[3], Sarah Haigh[3], Nicholas H. Crisp[3], Vitor T.A. Oiko[3], Rachel E. Lyons[3], Stephen D. Worral[3], Sabrina Livadiotti[3], Claire Huyton[3], Luciana A. Sinpetru[3], Rosa M. Domínguez[4], David González[4], Francesco Romano[5], Yung-An Chan[5], Adam Boxberger[5], Stefanos Fasoulas[5], Constantin Traub[5], Victor Jungnell[8], Kristian Bay[8], Jonas Morsbøl[8], Ameli Schwalber[9] and Barbara Heißerer[9]

1 UPC-BarcelonaTECH ESEIAAT, Terrassa, Barcelona, Spain

2 Euroconsult, Paris, France

3 The University of Manchester, Manchester, United Kingdom

4 Elecnor Deimos Satellite Systems, Puertollano, Spain

5 Institute of Space System, University of Stuttgart, Stuttgart, Germany

6 Mullard Space Science Laboratory (UCL), Dorking, United Kingdom

7 Christopher Newport University, Newport News, VA, United States of America

8 Gomspace AS, Aalborg East, Denmark

9 Concentris Research Management GmbH, Fürstenfeldbruck, Germany

*Address all correspondence to: daniel.garcia@upc.edu

IntechOpen

© 2020 The Author(s). Licensee IntechOpen. This chapter is distributed under the terms of the Creative Commons Attribution License (http://creativecommons.org/licenses/by/3.0), which permits unrestricted use, distribution, and reproduction in any medium, provided the original work is properly cited.

References

[1] Perez S et al. Prospects of Earth Observation Overview. Euroconsult. DISCOVERER H2020 Deliverable; 2018

[2] Denis G, Claverie A, Pasco X, Darnis J, de Maupeou B, Lafaye M, et al. Towards disruptions in Earth observation? New Earth Observation systems and markets evolution: Possible scenarios and impacts. Acta Astronautica. 2017;**137**:415-433. DOI: 10.1016/j.actaastro.2017.04.034

[3] UCS Satellite Database [Internet]. Union of Concerned Scientists. 2019. Available from: https://www.ucsusa.org/nuclear-weapons/space-weapons/satellite-database [Accessed: 13 May 2019]

[4] Kramer H. Observation of the Earth and Its Environment. Berlin: Springer; 2002

[5] Summerer L. Signs of potentially disruptive innovation in the space sector. International Journal of Innovation Science. 2011;**3**(3):127-140. DOI: 10.1260/1757-2223.3.3.127

[6] Christensen C. The Innovator's Dilemma. New York, NY: Harper Business; 2011

[7] Christensen C, Raynor M. The Innovator's Solution: Creating and Sustaining Successful Growth. 1st ed. Boston: Harvard Business Review Press; 2013

[8] Nagy D, Schuessler J, Dubinsky A. Defining and identifying disruptive innovations. Industrial Marketing Management. 2016;**57**:119-126. DOI: 10.1016/j.indmarman.2015.11.017

[9] Rogers E. Diffusion of Innovations. 4th ed. New York: Free Press; 1995

[10] Swanson E. Information systems innovation among organizations. Management Science. 1994;**40**(9):1069-1092. DOI: 10.1287/mnsc.40.9.1069

[11] Merges R, Reynolds G. The proper scope of the copyright and patent power. Harvard Journal on Legislation. 2000;**37**:45

[12] Abernathy WJ, Utterback JM. Patterns of industrial innovation. Technology Review. 1978;**80**:40-47

[13] Anderson P, Tushman M. Technological discontinuities and dominant designs: A cyclical model of technological change. Administrative Science Quarterly. 1990;**35**(4):604. DOI: 10.2307/2393511

[14] Dahlin K, Behrens D. When is an invention really radical? Research Policy. 2005;**34**(5):717-737. DOI: 10.1016/j.respol.2005.03.009

[15] Nagendra N, Segert T. Challenges for new space commercial earth observation small satellites. New Space. 2017;**5**(4):238-243. DOI: 10.1089/space.2017.0014

[16] EO Portal Directory [Online]. EO Sharing Earth Observation Resources. Available from: https://directory.eoportal.org/web/eoportal/satellite missions [Accessed: 19 May 2019]

[17] Planet SA [Online]. California: Planet Labs. 2019. Available from: https://www.planet.com/ [Accessed: 19 May 2019]

[18] Digital Globe SA [Online]. Westminster: DigitalGlobe Headquarters. 2019. Available from: https://www.digitalglobe.com/ [Accessed: 19 May 2019]

[19] Satellite Imaging Corporation [Online]. Houston: Satellite Imaging Corporation Headquarters. 2019. Available from: https://www.satimagingcorp.com/ [Accessed: 19 May 2019]

Chapter 8

A Survey on Small Satellite Technologies and Space Missions for Geodetic Applications

Vaios Lappas and Vassilis Kostopoulos

Abstract

Advances in microelectronics, materials, combined with affordable and frequent launch opportunities has led to a revolution which consists of small satellite missions used for technology validation, Earth observation, space exploration. Small satellites are now being developed in large volumes for mega-constellations for Earth observation, Internet of Things (IoT) and low latency communications (internet) thus democratizing space and making new space applications a reality. Advances in small satellite platforms, miniaturization of instruments and the availability of low-cost launches for small satellites, can enable new, geodetic missions which can benefit from the use of constellations of small satellites. An overview of some of the most important small satellite based geodetic missions is presented, along with a brief overview of new mission concepts which can significantly enhance our knowledge in the geodetic field.

Keywords: small satellites, GPS, geodetic, space missions

1. Introduction

Today, space geodetic techniques are the primary tools to study size, figure and deformation of the Earth, and its motion as a finite body in the inertial reference system. Space geodetic techniques have become the fundamental tools for geodesy, geodetic astronomy, and geodynamics. The development of space geodesy has increase significantly with the progress in space technology, miniaturization of satellites and the advent of Global Navigation Satellite Systems (GNSS) and GNSS technology such as receivers. GNSS is making a much greater impact in the last decade with the advent of small satellites, launch availability and the miniaturization of electronics. It should be viewed as the replacement of classical navigation and positioning (based on the observation of astrometric positions of natural celestial objects) by measurements of microwave signals emitted by artificial satellites [1–4].

With respect to space technology, the two main areas of geodetic research and space missions are:

- Gravity space missions: For geodesy and geodynamics, the CHAMP (Challenging Minisatellite Payload for Geophysical Research and Application) and CHAMP-FO missions, the GRACE (Gravity Recovery and Climate Experiment) mission, and the GOCE (Gravity field and Ocean Current

IntechOpen

Explorer) mission are particularly fascinating. Our knowledge of the Earth's gravity field (thanks to the use of space-borne GPS receivers, accelerometers, and gradiometers) has significantly grown thanks to these missions. Gravity missions are of central importance for altimetry, because the precise geoids are required to refer the sea surface topography to the geoid [1].

- GNSS. GNSS stands for Global Navigation Satellite System. The current generation of GNSS may be viewed as the successor of the Doppler systems. The systems are based on coherent microwave signals (in the L-band) emitted by the satellites in (at least) two carrier frequencies. Simultaneity of measurement of the signals emitted by several satellites and recorded by a receiver allow for instantaneous positioning. The GPS (Global Positioning System) is probably the best known GNSS and is considered the best-known space geodetic technique today. The system has an impact on science and society as a whole, reaching far beyond space geodesy. GPS revolutionized surveying, timing, pedestrian, car, marine and aircraft navigation. Many millions of receivers are in use today. Space-borne applications of the GPS have a deep impact on geodesy and atmospheric sciences [1]. The development new GNSS systems from Europe (Galileo), China (Beidou) and others will bring new developments and substantially improve our knowledge of Earth's and planetary geodesy.

In summary, satellite geodetic missions offer the following capabilities:

1. Provide precise relative and geocentric locations of separated points on Earth by ground to satellite measurements

2. Provide knowledge of the time invariant and time variant gravitational forces and surface forces acting on satellite from precise analysis of orbital perturbations and by satellite borne gravity scanning devices

3. Provide precise geocentric location of a spacecraft

4. Provide measurements of time invariant and time variant aspects of vertical geometry of the oceans and land on a rapid and continuing basis by satellite-borne altimeters

The advent of microelectronics, miniaturization in combination with the proven use of commercial off the shelf (COTS) electronics, has led to the wide use of small satellites, i.e., spacecraft which have a mass of 500–1000 kg. Small satellites have proved to be an affordable means of demonstrating new technologies, but also allowing constellations of satellites to approach conventional space missions in a different manner by increasing the revisit time over multiple locations of interest, increase the number of scientific measurements but also enable the use of differential measurements using ranging techniques in combination with space-borne GPS receivers. The next section presents an introduction to small satellites followed by an overview of the most important geodetic small satellite missions in orbit or completed to date.

2. Small satellites

The size and cost of spacecraft vary depending on the application; some can hold in your hand while others like Hubble are as big as a school bus. Small spacecraft (smallsats) focus on spacecraft with a mass less than 180 kg and about the size of a large kitchen fridge. Even with small spacecraft, there is a large variety of size and mass that can be differentiated. The accepted small satellite classification (per kg) in the space community is as follows:

- Minisatellite: 150–500 kg
- Microsatellite: 10–150 kg
- Nanosatellite: 1–10 kg
- Picosatellite: 0.01–1 kg
- Femtosatellite: 0.001–0.01 kg

Small satellites, miniaturized satellites, or smallsats, are satellites of low mass and size, usually under 500 kg (1100 lb). While all such satellites can be referred to as “small”, different classifications are used to categorize them based on mass. Satellites can be built small to reduce the large economic cost of launch vehicles and the costs associated with construction. Miniature satellites, especially in large numbers, may be more useful than fewer, larger ones for some purposes—for example, gathering of scientific data and radio relay. Small satellites have become a significant part of the space industry. The advent of small satellite technology in multiple areas such as micro-propulsion, cubesats, microelectronics, long distance communications and increased low cost rideshare launch availability has also led to new mission concepts and missions such as broadband internet, communications and Earth observation, which consist of mega constellations of small satellites. Given the available production capabilities of the Contractor and the trend of small satellites, seeking a demonstrator application in this market would be reasonable and would create market potential in the near future. The advent of microelectronics, electric micro propulsion and other small satellite subsystems has enabled a multitude of interplanetary and high power cubesat/nanosatellite mission to be

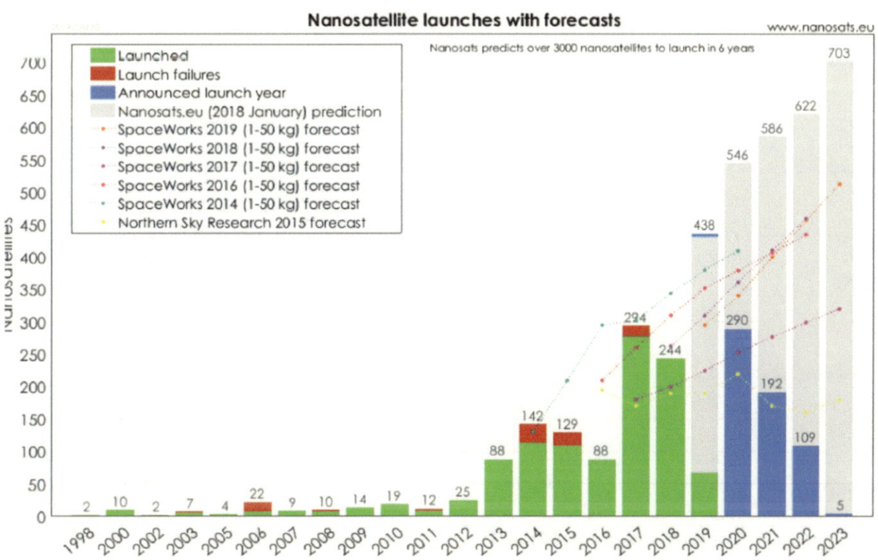

Figure 1.
Nano/microsatellite (<50 kg) forecast.

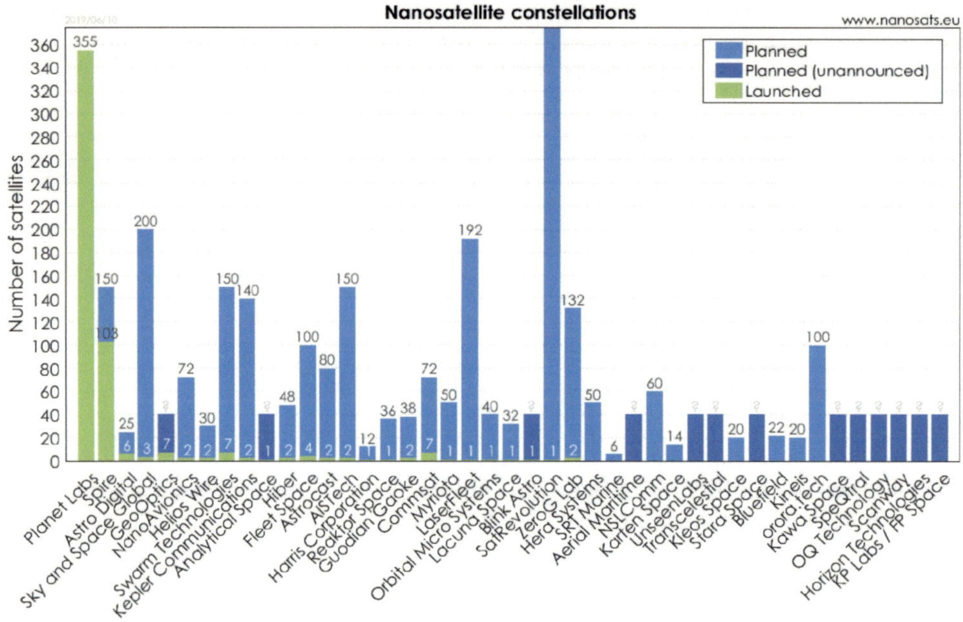

Figure 2.
Microsatellite (<50 kg) constellations launched and planned.

Con/n	Num.	Maker	Weight	Year	Avail	Orbit (km)	Status
Iridium Next	66 + 9 spares	Thales Alenia+ Orbital ATK	860 kg	2009	2018	780	Complete
Boeing	1396-2956	Boeing Satellite	N/A	2016	N/A	1200	Unknown
LeoSat	78-108	Thales Alenia	1250 kg	2015	2022	1400	first launches in 2021
OneWeb	882-1980	OneWeb Airbus JV	145 kg	2015	2020	1200	6 pilot satellites in February 2019
SpaceX Starlink	4425-11,943	SpaceX	227 kg	2015	2020	550 km	120 launched in 2020
O3b (SES)	20 O3b 7 O3bm	Thales Alenia (O3b) Boeing (O3bm)	700 kg	2008: O3b 2017: O3bm	2014: O3b 2021: O3bm	8000	O3b complete
Telesat LEO	117-512	TBC	300 kg	2016	2021	1000–1248	Prototypes 2018
CASIC Hongyun	156	CASC	<500 kg	2017	2022	160–2000	Prototypes 2018
CASC Hongyan	320	CASC	< 500 kg	2017	2023	1100	Prototypes l 2018

Table 1.
Mega-constellations planned/operations (assembled from multiple sources).

DOI: http://dx.doi.org/10.5772/intechopen.92625

Parameter	Airbus	OHB	SSTL	16 U	SITAEL
					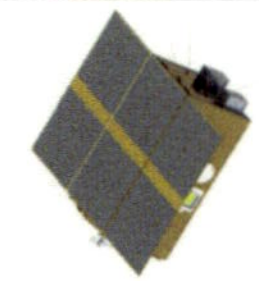
Platform mass (kg)	150	120	100	20	55
Payload mass (kg)	60	80	50	8	20
Dimensions (m)	1 × 1 × 1.3	0.8 × 0.845 × 0.8	0.7 × 0.45 × 0.77	0.3 × 0.04 × 0.4	0.32 × 0.32 × 0.4
Propulsion	Xenon HET	Multiple	Resistojet	Multiple	Multiple
Lifetime (years)	>5	>5	5	5	3

Table 2.
Small satellite comparison (assembled from multiple sources).

developed. JPL's MaRCO 3 U cubesat mission are operating already in Mars orbit as communication relays and the NASA funded LunarCube is a 100 W 3 U cubesat mission which will use a Busek iodine RF thruster for Lunar Exploration.

2.1 Small satellite market and trends

In addition to cubesats, larger small satellite missions are now being designed for high power missions due to the sophistication of their payloads and need to use micro electric propulsion which drive power demands. The smallsat (<500 kg mass satellites) market is going through significant expansion in terms of capabilities and demand. In the last couple of years, numerous companies have produced solutions, largely based around a constellation approach, to better deliver services and reach out to new users. These solutions are supported by new ventures, entrepreneurs investing in space in this so-called new or adaptive space environment. The logic of lower-cost constellations is to provide global connectivity from one system (satcom) or high-frequency change detection (in Earth observation [EO]) or connecting devices and vehicles (Information) for the Internet of Things (IoT), Machine to Machine (M2M), and traffic monitoring (AIS or ADS-B). It is aided by the advancement of satellite system miniaturization permitted by new technologies and/or advances in related sectors, particularly in computational technology with smallsats now providing operational services that were previously only achievable through heavier satellites. Over 300 nano/microsatellites were launched in 2017, shattering analyst' expectations and surpassing even SpaceWorks "full market potential" prediction from last year. 2017 represented a 205% increase in nano/microsatellites launched compared to 2016. SpaceWorks' estimates up to 2600 nano/microsatellites will require launch over the next 5 years. This creates space for new opportunities that could benefit from the technology developments in the activity and match the size and scope of the involved partners.

As it can be seen in **Figures 1**, **2** and **Table 1**, mega-constellations with microsatellites up to a mass of 200–300 kg are the most popular trend in the space industry and indicate that the 200–300 kg mass and volume factor of 0.125–1 m^3 are the near optimum to: (i) squeeze as many small satellites in medium size launch vehicles such as Falcon 9, Ariane 6 (ii) still launch 2–3 small satellites in dedicated small launchers VEGA, LauncherOne, Electron etc. (iii) are small enough to mass produce in lean, highly automated, automotive style production lines as OneWeb and SpaceX are currently undertaking.[1]

Table 2 presents a summary and comparison of a representative selection of small satellites of different size, country of manufacture and capabilities.

3. Small satellite geodetic missions

Small satellites have significant benefits which have made them very popular in the last decade for scientific and commercial space missions:

- The advent of microelectronics, automated manufacturing, sensors, batteries, motors and the significant reduction in their cost of manufacture has enabled the miniaturization of larger conventional satellites and instruments brining the launch cost (cost per kg) significantly down

1 https://www.nanosats.eu/#figures

- Use of COTS components for Low Earth Orbit (LEO) small satellite missions has also lowered space mission costs, helped towards miniaturization and also has created large availability of components and sensors which can be used for space instruments, subsystems and platforms

- Miniaturized satellites such as cubesats or nano/microsatellites can be launched as secondary payloads which are very affordable

- New private and civilian launchers have increased the number and competitiveness of launches, bringing launch costs down and increasing launch availability significantly compared to 10 years ago

- Access to private funding, entrepreneurship has increased small satellite subsystem, sensor and platform quantities and thus reducing mission costs

- Miniature small satellites with low launch costs can enable constellations which can thus allow for a tremendous increase in the number, frequency and location of sensor measurements simultaneously across the globe

All the above advantages of small satellites have contributed to new space missions which were simply impossible, or too expensive to achieve in the past. Specifically for geodesy, there have been a number of scientifically rich satellite missions which used radar and laser altimetry as well as geodetical components to study Earth's gravity field such as CHAMP (CHAllenging Mini satellite Payload), GRACE (Gravity Recovery and Climate Experiment), and GOCE (Gravity field and steady-state Ocean Circulation Explorer). These satellite missions, although in the boundaries of small satellite missions (with respect to mass and cost) used small satellite technologies to achieve complex mission objectives and provide unique science data about Earth's gravity field.

Table 3 shows a summary of small satellite geodesy mission types. In summary, the benefits of small satellites geodetic missions are that they can be a cost effective means to measure/estimate Earth's geopotential using ranging and GPS data, while constellations can help get polar region data (missing), increase temporal frequency (from months to weeks) and increase the amount, frequency and location of science measurements at a reasonable mission cost. The following section describes the most significant geodetic small satellite missions launched to date.

3.1 Reflector type satellite missions

GFZ 1 (Geo Forschungs Zentrum Potsdam 1) is the first satellite mission designed and funded by the GeoForschungsZentrum Potsdam, Germany. The mission objectives of GFZ-1 were to determine variations in the rotational characteristics of the Earth and to measurement changes in the Earth's gravity field. For the high-resolution determination of the parameters of the gravity field the satellite had to be launched into the lowest possible orbit. At its altitude of 400 km, GFZ-1 was the lowest geodynamic satellite to be ranged to by lasers (in 1995). As the vehicle's orbit was decaying, the satellite's orbital motion was used to calculate atmospheric densities [5, 6] (**Figure 3**).

GFZ-1 was s a passive geodetic satellite which only used one instrument, the retroreflector array (RRA). The GFZ-1 RRA consisted of 60 corner cubes and has a center of mass correction of 58 + 2 mm. These retroreflectors were quartz prisms placed in special holders recessed into the satellite's body. External metallic surfaces were covered with white paint for thermal control purposes and to facilitate visual

Mission	Approach	Example	Notes
Geodesy: measurement of the Earth's gravitation field	1 satellite with accelerometer + GPS GPS: m accuracy Accelerometer: 10^{-10} m/s^2 (drag 0^{-6} m/s^2)	CHAMP (2000)	Low cost mission with medium resolution data
Geodesy with retroreflector array	1 satellite with retroreflector array	GFZ-1, Larets	Low resolution data
Geodesy by Differential measurement	2 satellites with accelerometers, GPS, ranging Ranging: nm to μm	GRACE (2002) GRACE-FO (2010)	Performance limited due to acceleration noise
Multipoint geodesy (constellation), Reflectometry	Satellites with accelerometers, GPS, ranging Ranging: nm to μm	TDS-1, GYCNSS, C-2	Application to multiple fields
Drag-free based geodesy	2 drag-free satellites with ranging, GPS Accuracy: 10^{-10} to 10^{-15} m/s^2 Drag force canceled	Cubesats	Need high precision attitude/ orbit determination/control

Table 3.
Small satellite geodetic mission types with mission examples.

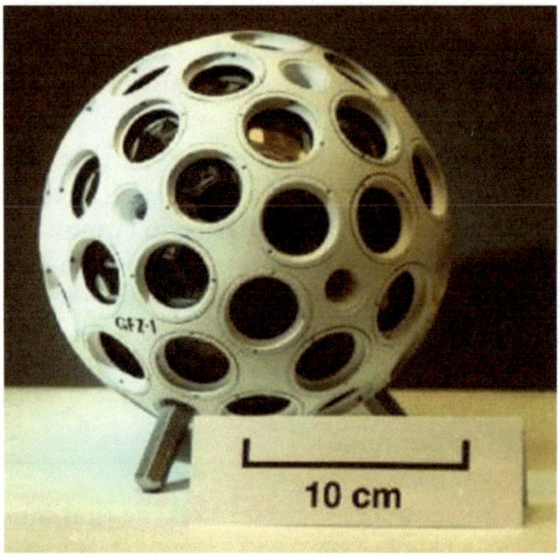

Figure 3.
Model of the geo research satellite GFZ-1 (GFZ) [6].

observation in space. The satellite was built by Kayser Threde GmbH and launched through the space station MIR bby RKK Energiya and had a mass of 21 kg. GFZ-1 was transported to Mir Station aboard the Russian Progress-M 27 spacecraft and from there put into a low Earth orbit in April 1995 at an orbit of 382 km × 395 km with an inclination 51.6°. On June 23rd, 1999, GFZ-1 completed its mission with the satellite burning up in the Earth's upper atmosphere. The GFZ-1 orbited nearly 24,000 times around the Earth and for 4 years and 64 days in space, 5402 passes of GFZ-1 were observed by 33 cooperating ground stations using satellite laser ranging, around the globe [5, 6].

The LARES mission concept represents an improvement of the LAGEOS-3 project proposed in 1984 by Ciufolini. The LAGEOS (Laser Geodynamics Satellite) series was designed to be a passive long-lived satellite with a stable, well-defined orbit. As such, it acts as a reference point in inertial space. An international ground-based network of laser ranging stations used the orbiting LAGEOS satellites as passive reflectors to obtain ranges to the satellite by precision laser echo-bounce techniques. Since the position of the satellites is determined by some laser ranging stations with uncertainties of <1 cm, there was a potential for measuring the 2 m per year drift of the nodes of LAGEOS satellites in orbit. The LAGEOS-1 (LAser GEOdynamics Satellite-1, launch May 4, 1976) and LAGEOS-2 (launch Oct. 23, 1992) missions in MEO (Medium Earth Orbit) of NASA and ASI represented the origin of international cooperative research in geodynamics [7, 8] (**Figure 4**).

LARES was a completely passive satellite made of a dense tungsten alloy (THA-18 N) sphere of 376 mm in diameter and a mass of ~400 kg (density of ~18 kg/cm^3) covered with retroreflectors that allowed the satellite's motion to be followed via SLR (Satellite Laser Ranging) from Earth. Once in orbit, LARES became the satellite with the highest mean density in the Solar System. The surface of the sphere was covered by 92 CCRs (Corner Cube Reflectors) evenly distributed so that the signal strength is practically independent on satellite attitude. The LARES spacecraft was launched on February 13, 2012 on the maiden flight of the Vega launch vehicle of ESA (the Vega flight was designated as VV01); the launch site was Kourou in French Guiana [7, 8].

Larets was a small geodesy and calibration satellite launched for IPIE (Institute of Space Device Engineering—Moscow). It was a 21 cm sphere with 60 laser retro-reflectors, very similar to the German GFZ 1 satellite. Larets was designed to address scientific and applied problems in the interests of geodesy and geodynamics. The LARETS mission launched in 1003 with a 21 kg mass and in a 675 km orbit, was a next generation satellite based on a low-target-error laser satellite design optimization, which started with the WESTPAC mission, another reflector type satellite mission launched in 1998 in a 835 km, 98.67°. Larets was launched with a cluster of small satellites in September 2003 on a Kosmos-3M rocket [9] (**Figure 5**).

3.2 Geodesy with GPS measurements of the Earth's gravitation field

Challenging Minisatellite Payload (CHAMP) was a German satellite launched July 15, 2000 from Plesetsk, Russia and was used for atmospheric and ionospheric research, as well as other geoscientific applications, such as GPS radio

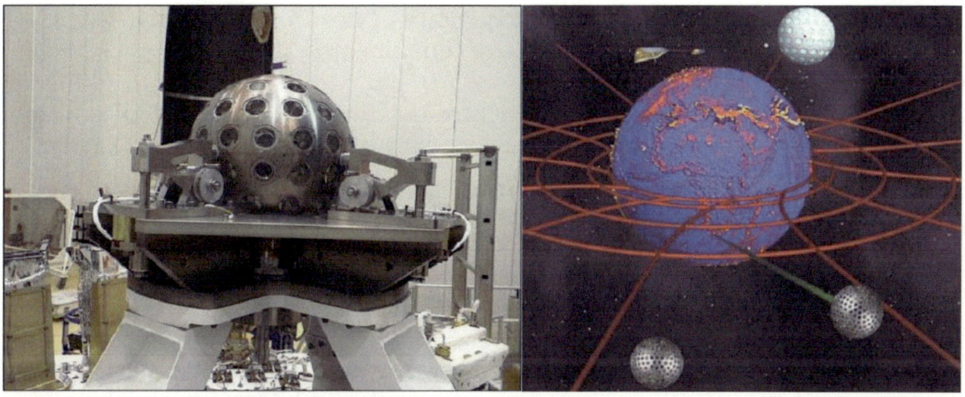

Figure 4.
(Left) Image of LARES satellite (Right) Schematic view of LAGEOS-1, LAGEOS-2 and LARES in orbit for the measurement of frame dragging [8].

Figure 5.
Larets satellite model [9].

occultation. CHAMP was managed by GeoForschungsZentrum (GFZ) Potsdam (GFZ). The spacecraft was the first application of Astrium's "Flexbus" platform (now Airbus Defense and Space); GRACE was the second. A heavily modified version flew as the GOCE mission. CHAMP completed its mission and re-entered the Earth's atmosphere on 19 September 2010 after 10 years (design life: 5 years) [10]. The three primary science objectives of the CHAMP mission were to provide:

- highly precise global long-wavelength features of the static Earth gravity field and the temporal variation of this field.
- high accuracy global estimates of the main and crustal magnetic field of the Earth and the space/time variability of these field components
- large number of GPS signal refraction data, with a good global distribution, caused by the atmosphere and ionosphere, which can be converted into temperature, water vapor and electron content.

With its multifunctional and complementary payload CHAMP aimed at contributing to the following Earth system components:

- Geosphere: investigation of the structure and dynamics of the solid Earth from the core along the mantle to the crust, and investigation of interactions with the ocean and atmosphere
- Hydrosphere: more accurate monitoring of ocean circulation, global sea level changes and short-term changes in the global water balance as well as interactions with weather and climate
- Atmosphere: global sounding of the vertical layers of the neutral and ionized gas shell of the Earth and relationship with weather on Earth and space weather (**Figure 6**).

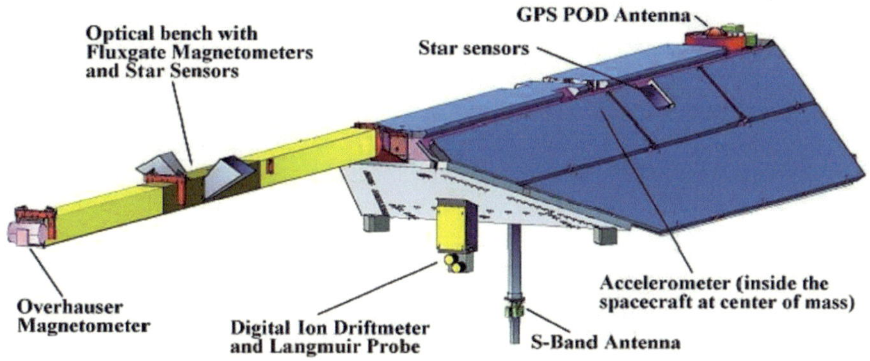

Figure 6.
Front view of the CHAMP spacecraft (image credit: GFZ).

The CHAMP satellite had a robust structure design with fixed solar panels. The primary structure was mainly based on aluminum sandwich panels with an additional kapton foam layer on the outer panels. The shape of the satellite was a compromise with respect to its aerodynamic behaviour, accommodation of instruments and subsystems and fitting into the fairing of the launcher [10].

- Total mass: 522 kg
- Height: 750 mm
- Length (with 4044 mm Boom): 8333 mm
- Width: 1621 mm
- Area to mass ratio: 0.00138 m^2/kg

3.3 Multipoint geodesy and Reflectometry

GRACE was an international cooperative US-German dual-minisatellite SST (Satellite-to-Satellite Tracking) geodetic mission with the overall objective to obtain long-term data with unprecedented accuracy for global (high-resolution) models of the mean and the time-variable components of the Earth's gravity field (a new model of the Earth's gravity field every 30 days for 5 years). GRACE was also part of NASA's ESSP (Earth System Science Pathfinder) program [11, 12]. The science objectives were:

- To enable a better understanding of ocean surface currents and ocean heat transport
- To measure changes in the sea-floor pressure
- To study ocean mass changes
- To measure the mass balance of ice sheets and glaciers
- To monitor changes in the storage of water and snow on the continents

The mission concept made use of measurements of the inter-satellite range changes and its derivatives between two co-planar satellites (in low-altitude and polar orbits), using a microwave tracking system. The orbits of the two separately flying satellites are perturbed differently in the Earth's gravity field, leading to inter-satellite range variations. In addition, each satellite carried a GPS receiver of geodetic quality and high-accuracy accelerometers to enable accurate orbit determination, spatial registration of gravity data and the estimation of gravity field models. The fluctuations in the strength of the Earth's gravity field reflect in turn changes in the distribution of mass in the ocean, atmosphere, and solid Earth, and in the storage of water, snow, and ice on land. Since ocean bottom pressure represents a column integral of the mass of the atmosphere plus ocean, this measurement technique permits the deduction of ocean bottom pressure changes from space [11, 12].

Both satellite structures were of identical design. The shape of each satellite is trapezoidal in cross section, based on the FLEXBUS design of Astrium (now Airbus Defense and Space) (length = 3122 mm, height = 720 mm, bottom width = 1942 mm, top width = 693 mm). The FLEXBUS structure consists of CFRP (Carbon Fiber Reinforced Plastic). This material, with a very low coefficient of thermal expansion, provides the dimensional stability necessary for precise range change measurements between the two spacecraft. The actuators include a cold gas system (with 12 attitude control thrusters and two orbit control thrusters, each rated at 40 mN) and three magnetorquers [11, 12] (**Figure 7**).

Each satellite has a mass of 432 kg (science payload = 40 kg, fuel = 34 kg) and the satellite power is 150-210 W (science payload = 75 W). The top and side panels of each satellite are covered with strings of silicon solar cells; NiH batteries with 16 Ah provide power storage. The satellite design life was 5 years. About 80% of the spacecraft's on-board electronics parts were COTS (Commercial Off-the-Shelf) products. Mission operations exceeded the initial 5 years, operating for 15 years until the decommissioning of GRACE-2 on 27 October 2017 [11, 12] (**Figure 8**).

Based on the significant success of GRACE, the GRACE-FO mission, is also a collaboration between NASA and GFZ and was launched on 22 May 2018 aboard a SpaceX Falcon 9 rocket from Vandenberg AFB, California, sharing the launch with five Iridium NEXT satellites [13]. During in-orbit checks, an anomaly was discovered in the primary system component of the microwave instrument (MWI), and the system was temporarily powered down on 19 July 2018. After a full investigation by an anomaly response team at JPL, the backup system in the MWI was powered up on 19 October 2018 and GRACE-FO resumed its in-orbit checks. GRACE-FO entered the science phase of its mission on 28 January 2019 [13]. The orbit and design of GRACE-FO is very similar to its predecessor. GRACE-FO has a design life of 5 years. Astrium (now Airbus DS) uses a 3rd generation Flexbus for the GRACE-FO mission. Each of the GRACE-FO satellites measures approximately

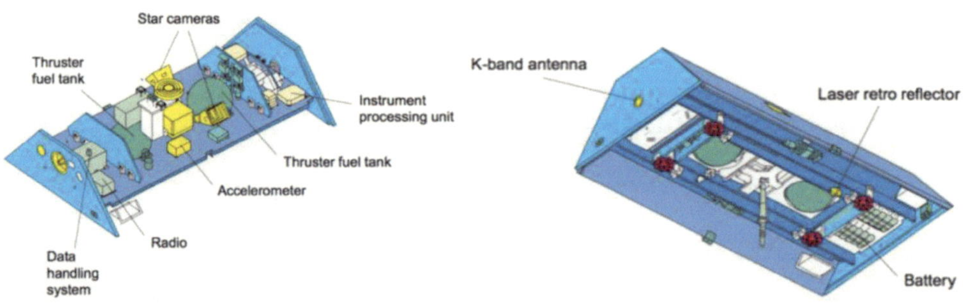

Figure 7.
GRACE satellite design [11, 12].

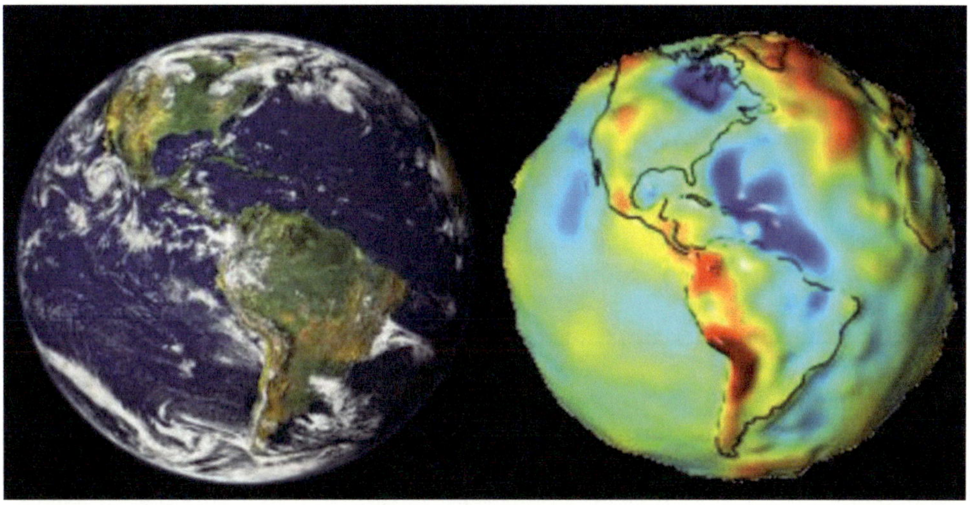

Figure 8.
GRACE gravity field maps [11, 12].

3 m × 2 m × 0.8 m and has a mass of around 580 kg. GRACE-FO employs the same two-way microwave-ranging link as GRACE, which will allow for similar inter-satellite ranging precision (**Figure 9**).

In addition, GRACE-FO employs laser-ranging interferometry (LRI) as a technological experiment in preparation for future satellites. The LRI allows for more accurate inter-satellite ranging due to the shorter wavelength of light, and additionally allows the angle between the two spacecraft to be measured as well as their separation, via differential wavefront sensing. Using the LRI, scientists have improved the precision of the separation distance measurements by a factor of more than 20 relative to the GRACE mission. The LRL lasers must be detected by a spacecraft about 137 miles (220 km) away. This laser approach will generate much more accurate measurements than the previous GRACE satellite mission. The GRACE-FO satellites obtain electricity from gallium arsenide solar cell array panels covering the outside of each satellite. GRACE-FO will continue to monitor Earth's gravity and climate. The mission will track gravitational changes in global sea levels, glaciers, and ice sheets, as well as large lake and river water levels, and soil moisture. In addition, each of the satellites will use GPS antennas to create at least 200 profiles per day of atmospheric temperature distribution and water vapor content, a first for the GRACE mission [13].

Figure 9.
(Left) Artist concept of GRACE-FO Mission; (Right) the gravity recovery and climate experiment follow-on (GRACE-FO) twin satellites [13].

3.3.1 GNSS reflectometry

GNSS-Reflectometry is a new technique that shows promise for many Earth observation applications including remote sensing of oceans, land, and ice. High grade GNSS payload has been developed in the last few years, that are low in size and power, and suitable for use on small satellites. One such GNSS payload is the SGR-ReSI GNSS Reflectometry Instrument flown on the TechDemoSat-1 microsatellite mission, launched in July 2014. The instrument has been operational since its commissioning in September 2014 and has been collecting delay Doppler maps routinely over many different surfaces. Preliminary work has been undertaken to develop and validate wind speed inversion algorithms against Advanced Scatterometer (ASCAT) wind measurements with promising results [14]. One of the first ever GPS Reflectometry experiments was flown again by Surrey Satellite Technology Limited on the UK-DMC microsatellite mission, launched in 2003, which proved the feasibility of using GNSS reflections for measuring the sea state and other geophysical observables through a partnership between SSTL and National Oceanography Centre (NOC) [14]. Preliminary work using UK-DMC GPS data focused on inverting the measurements into Level 2 products, specifically wind speed and mean-squared slope over the ocean, with promising results. Reflections recovered over the land surface showed a strong geophysical imprint, suggesting potential for hydrological and vegetation related retrieval. Subsequently, a new instrument was developed called the Space GNSS Receiver Remote Sensing Instrument (SGR-ReSI) to gather more space-borne reflectometry data and demonstrate the potential for a sea-state service. In parallel, a US mission called CYGNSS was selected by NASA that plans to measure hurricanes with reflected GNSS signals collected using an updated revision of the SGR-ReSI, (also referred to as delay Doppler mapping Instrument) as payload on each of the eight satellites [14, 15] (**Figures 10** and **11**).

3.3.2 Cyclone Global Navigation Satellite System (CYGNSS)

The Cyclone Global Navigation Satellite System (CYGNSS) is a space-based system developed by the University of Michigan and Southwest Research Institute with the aim of improving hurricane forecasting by better understanding the interactions between the sea and the air near the core of a storm. CYGNSS estimates the wind speed from its radar measurements [15–17]. Winds are measured continuously over the ocean in all weather conditions, although the mission objectives are focused on measurements made in and near the inner core of tropical cyclones Each CYGNSS satellite (eight in total) carries a Delay Doppler Mapping Instrument (DDMI), consisting of:

- a Delay Mapping Receiver (DMR)
- two nadir-pointing antennas
- one zenith-pointing antenna

The instrument receives GPS signals scattered by the ocean surface for the purposes of bi-static scatterometry. The CYGNSS mission was launched on December 15, 2016, at 13:37:21 UTC from a single Pegasus XL air-launched rocket at 35° inclination and 520-km altitude orbit. The eight CYGNSS microsatellites include a Delay Doppler Mapping Instrument consisting of a multichannel GPS receiver, low gain zenith antennas and high gain nadir antenna.

Figure 10.
(Left) TechDemoSat-1 (TDS-1) prior to launch (Centre) SGR-ReSI SGR-ReSI, part of sea-state payload on TDS-1 (Right) TDS-1 nadir GNSS antenna [14].

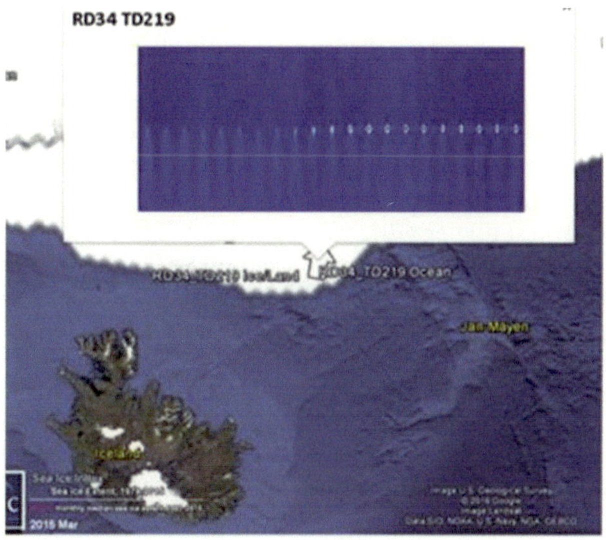

Figure 11.
Sequence of delay Doppler measurements (DDMs) over a sea/ice boundary near Iceland, march 2015. The white arrow indicates the boundary as detected by canny edge detection routine [14].

Attitude is three-axis stabilized with 2.1° (3σ) knowledge and 2.8° (3σ) control using horizon sensors, a magnetometer, pitch momentum wheel, and torque rods. Satellite mass and power are estimated to be ~25 kg and ~38 Watts. The satellites were built by SwRI and the payload by SSTL USA, based on SSTL's TDS-1 GPS payload [14, 16] (**Figure 12**).

Figure 12.
(Left) CYGNSS microsatellite platform built by SwRI (Right) delay Doppler mapping instrument (DDMI).

3.3.3 Formosa Satellite-7/Constellation Observing System for Meteorology Ionosphere and Climate (FORMOSAT-7/COSMIC-2)

The Formosa Satellite-7/Constellation Observing System for Meteorology Ionosphere and Climate (FORMOSAT-7/COSMIC-2, hereafter C2), is a recently launched equatorial constellation of six satellites carrying advanced radio occultation receivers, which exhibit high signal-to-noise ratio, precision, and accuracy, and the ability to provide high vertical resolution profiles of bending angles and refractivity, which contain information on temperature and water vapor in the challenging tropical atmosphere. The mission is an international collaboration between Taiwan (NSPO) and the United States (NOAA) that will use a constellation of 12 remote sensing microsatellites (planned) to collect atmospheric data for weather prediction and for ionosphere, climate and gravity research. Budget constraints have meant that the constellation will most likely remain to the current number of 6 currently in orbit [17]. The first six satellites, built by SSTL (UK) were launched on June 25 via the SpaceX Falcon 9 launcher to an initial circular parking orbit of 720 km. Eventually, they were positioned in a low inclination orbit at a nominal altitude of ~520–550 km with an inclination of 24° (using their propulsion system). Through constellation deployment, they are placed into 6 orbital planes with 60° separation. The satellites, built by SSTL in the UK have a mass of 300 kg and are based on SSTL's 150 kg microsatellite platform, with dimensions 100 × 125 × 125 cm^3, uses S band (2 Mbps download), 2 GB data storage, utilize nine downlink stations and have a hydrazine mono-propellant propulsion system to lower their initial 725 km orbit to a 550 km final operational orbit.

After an initial calibration/validation phase, over 100,000 soundings of bending angles and refractivity that passed quality control in October 2019 and are being compared with independent data, including radiosondes, model forecasts, and analyses. The comparisons show that C2 data meet expectations of high accuracy, precision, and capability to detect super refraction. When fully operational, the C2 satellites are expected to produce ~5000 soundings per day, providing freely available observations that will enable improved forecasts of weather, including tropical cyclones, and weather, space weather, and climate research (**Figure 13**).

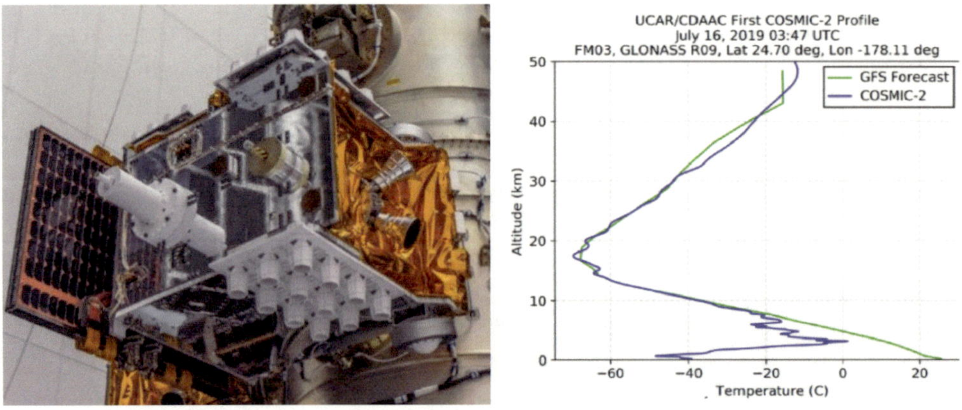

Figure 13.
(Left) COSMIC-2 on Falcon 9 Heavy (Right) 110 neutral atmospheric occultations within 3 h of receiving the first level 0 data [16].

3.4 Drag free space missions

Drag-free space systems [11–19] provide autonomous precision orbit determination, more accurately map the static and time varying components of Earth's mass distribution, aid in our understanding of the fundamental force of gravity, and ultimately open up a new window to our universe through the detection and observation of gravitational waves. At the heart of this technology is a gravitational reference sensor, which (a) contains and shields a free-floating test mass from all non-gravitational forces, and (b) precisely measures the position of the test mass inside the sensor. A feedback control system commands thrusters to fly the "tender" spacecraft with respect to the test mass [16]. Thus, both test mass and spacecraft follow a pure geodesic in spacetime. By tracking the position of a low Earth orbiting drag-free satellite we can directly determine the detailed shape of geodesics and through analysis, the higher order harmonics of the Earth's geopotential. In addition to geodesic information, the commanded thrust, test mass position and GPS tracking data can be combined to produce three dimensional maps of atmospheric winds and density. With multiple drag-free spacecraft, one can perform a more accurate differential measurement between two geodesics, for example with laser interferometry, in order to improve measurements made by NASA's twin GRACE satellites [12, 18].

The range of applications for drag-free technology is broad. A summary is provided in **Table 3** combined with other geodesic mission types [20]. The listed applications are separated into four distinct categories: navigation, Earth science, fundamental physics, and astrophysics. Two key performance metrics for each application are also shown. The first metric, called drag-free performance, is the residual acceleration of the test mass. For an ideal drag-free satellite, the residual acceleration is zero, but in practice small, residual forces act on the test mass, perturbing its trajectory with respect to a pure geodesic. The primary goal of drag-free satellite design is to minimize these residual forces. The second metric, called metrology, is either the measurement of the absolute position of a drag-free test mass (e.g. via GPS) or the differential measurement of the distance between two drag-free test masses. Space missions which implemented this concepts were the drag-free satellites: NASA's Gravity Probe B (GP-B), which tested two predictions of general relativity with ultra-precise drag-free gyroscopes in low Earth orbit [21–23], and ESA's geodesy mission, the Gravity field and steady-state Ocean Circulation Explorer (GOCE) [23].

As mentioned, the miniaturization of small satellites and instrumentation and the overall low cost of small satellite missions, makes them ideal vehicles to expand drag free mission principles. Section 4 details multiple cubesat based drag free mission concepts which are in the design phase and which will fly into space in the next few years. In the following section a brief overview of the GRACE and GOCE missions is presented.

4. Geodetic smallsat mission concepts

The availability of reusable launch vehicles such as the SpaceX Falcon 9 and 9 Heavy and the further miniaturization of small satellites and instruments to the cubesat form factor (1–3 U) and mass (<5 kg) has led to the proposal of multiple constellation based geodetic missions which are currently being investigated. The section below provides a brief overview of some of these mission concepts [19–27].

CNES has performed the preliminary design studies of a mission for a future nano-satellite constellation of GNSS-RO receivers for a targeted number of occultations:

- 10.000 per day main identified technical specifications to meet the end user requirements and comply with the low-cost constraint
- small size and small mass (<50 kg) with a minimalist instrumentation to remain a low-cost system
- main planned option for the constellation: 8 LEO satellites, altitude of 600 km
- receiver bi-frequencies (E1 and L5) and bi-GNSS constellation (GPS + Galileo)
- 2 polar ground stations
- array antenna with at least 8 dB gain and 50° coverage

MicroGEM (Microsatellites for GNSS Earth Monitoring) was a phase A study which proposed the use of satellites with a mass of 100 kg for monitoring of the Earth. The study explored how miniaturization of microsatellites and instrumentational can enable low cost geodetic missions. Using GPS/Galileo receivers and the GFZ-satellite missions CHAMP (CHAllenging Minisatellite Payload) and GRACE (Gravity Recovery and Climate Experiment) philosophy a small constellation is proposed. MicroGEM studied how to link the CHAMP and GRACE missions with cubesat/microsatellite technology and proposed for the first time to use the signals from the future Galileo-satellites for the GNSS-supported atmospheric and ionospheric remote sensing on a global scale. The study analyzed the significant improvements in this method. It also proposed to use GNSS-signals for the remote sensing of ocean and ice surface will be employed for the first time. The particular technological challenge of MicroGEM lies in the fact that this small satellite mission could serve as a predecessor for future multi-satellite systems with scientific GNSS-receivers as satellite payload. With such constellations the number of measurements can be considerably increased, and an improved global coverage can be achieved [24].

The proposed PRETTY (Passive Reflectometry and Dosimetry) mission included a demonstrator payload for passive reflectometry and scatterometry focusing on very low incidence angles whereby the direct and reflected signal will be received via the same antenna. The correlation of both signals will be done by a specific FPGA based hardware implementation. The demonstration of a passive reflectometer without the use of local code replica implicitly showed that also signals of unknown data modulation can be exploited for such a purpose. The PRETTY mission was proposed by an Austrian consortium with RUAG GmbH as prime contractor, relying on the results from a previous CubeSat mission (OPS-SAT) conducted by TU Graz under ESA contract [25].

GEOCON was another cubesat type study which is investigating the development of a new measurement concept using one or more space-based reference points (satellites) to significantly reduce the errors in the site ties between co-located geodetic ground stations. The proposed concept uses a novel idea of upconverting the Global Navigation Satellite System (GNSS) signal received at the satellite and transponding it to a Very Long Baseline Interferometry (VLBI) antenna ground station. This approach does not require the satellite to be in view of more than one VLBI station at a time, allowing the use of Low Earth Orbits. This is advantageous since it opens up the possibility of using inexpensive CubeSats or other small satellites, making it feasible to implement a cost-effective constellation of such satellites (GEOCON) to provide better global coverage and further improve the accuracy of the site ties for the Global Geodetic Observing System—GGOS, stations' network.

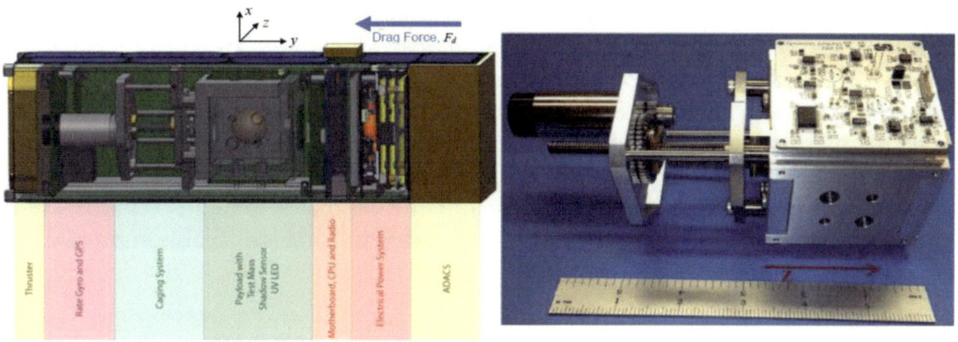

Figure 14.
CAD model of the Drag-free CubeSat (Right) drawing of the 1 U GRS payload [28].

A Drag-Free CubeSat mission has been proposed recently, to demonstrate the feasibility of a Gravitational Reference Sensor (GRS) with an optical readout for a 3 units (3 U) spacecraft [27, 28]. A purely drag-free object is defined by the absence of all external forces other than gravity, which are shielded by the spacecraft. In a real case, the spherical test mass (TM) will still be affected by disturbances. Several of them are passively reduced by the design of the TM housing. This system is a thick-walled aluminum box that holds the shadow sensors and shields the TM. The housing has an effect on the mechanical, thermal and magnetic environment around the TM. All of them have been analyzed. The mechanical vibrations have to fit the launch environment and the modes have to be outside of the measurement range (0.0001–1 Hz). The housing, together with the TM, the sensors and the UV LEDs for charging control, constitutes the GRS, which would then fit into a 1 U. The other 2Us are occupied by the caging mechanism that constraints the TM during launch, the thrusters, the Attitude Determination and Control System (ADACS) and the electronics. The Drag-Free CubeSat will be the result of the combined efforts of Stanford, University of Florida, KACST and NASA and will be the first drag-free mission with an optical readout and the first GRS designed within the limits of a 3 U small satellite [27].

The 1 U GRS consists of a 25 mm diameter spherical test mass housed inside a 50 mm cubic cavity. The sphere's position is sensed with a LED-based differential optical shadow sensor, its electric charge is controlled by photoemission using UV LEDs, and the spacecraft position is maintained with respect to the sphere using a cold gas micro-propulsion system. The Drag-free CubeSat is a 3-unit (3 U) CubeSat, measuring 34 cm × 10 cm × 10 cm and weighing 4 kg at launch. The drag-free control system uses the satellite position measurement provided by the shadow sensor and a small cold gas thruster in the aft of the satellite to compensate for atmospheric drag and keep the spacecraft centred with respect to the test mass. A commercially available Attitude Determination and Control System (ADACS) will maintain the satellite's attitude pointed in the direction of the drag force, as well as control the satellite's roll angle [28]. The target performance of the Drag-free CubeSat is roughly 10 times better than the GRACE accelerometers and comparable to the drag-free performance of GOCE. The performance is limited primarily by the minimum impulse bit and thrust noise of available CubeSat scale thrusters (**Figure 14**).

5. Conclusion

Geodetic missions have benefited from small satellite technology, with space missions such as GOCE, CHAMP, CRACE, GRACE-FO. Advances in GPS technology, sensors (accelerometers), microelectronics and the wide use of cubesat (<5 kg) nanosatellite technology allows for drastic mass and volume reduction of satellite platforms and instruments and thus allowing microsatellite/cubesat type constellations to offer unique measurement capabilities for geodetic applications. New mission concepts using reflectometry, occultation, ionospheric research and drag free payloads are presented focused on cubesat level technology and size factors, which in combination with lowering launch costs and an increase in launch opportunities bring a new setting to conduct unprecedented geodetic missions with higher temporal resolution and higher measurement accuracy compared to what is available at present.

Author details

Vaios Lappas* and Vassilis Kostopoulos
Applied Mechanics Lab, University of Patras, Rio, Greece

*Address all correspondence to: vlappas@upatras.gr

IntechOpen

© 2020 The Author(s). Licensee IntechOpen. This chapter is distributed under the terms of the Creative Commons Attribution License (http://creativecommons.org/licenses/by/3.0), which permits unrestricted use, distribution, and reproduction in any medium, provided the original work is properly cited.

References

[1] Beutler G. Revolution in Geodesy and Surveying. International Federation of Surveyors. Article of the Month: July 2004. Available from: https://www.fig.net/resources/monthly_articles/2004/july_2004/beutler_july_2004.pdf

[2] Potential Applications of Satellite Geodetic Techniques to Geosciences, by United States. National Aeronautics and Space Administration, Ad Hoc Advisory Group. NASA Report SP-158; 1968

[3] Heinkelmann R. VLBI Geodesy: Observations, Analysis and Results, Geodetic Sciences—Observations, Modeling and Applications. Rijeka, Croatia: IntechOpen; 2013. DOI: 10.5772/54446. Available from: https://www.intechopen.com/books/geodetic-sciences-observations-modeling-and-applications/vlbi-geodesy-observations-analysis-and-results

[4] Seeber G. Satellite Geodesy: Foundations, Methods, and Applications. De Gruyter Publishing; 2008. DOI: 10.1515/9783110200089

[5] Kramer HJ. Observation of the Earth and Its Environment: Survey of Missions and Sensors. Springer-Verlag Berlin Heidelberg: Springer; 2002. ISBN: 978-3-642-56294-5

[6] Website of GeoForschungs Zentrum Potsdam. Available from: https://www.gfz-potsdam.de/en/section/global-geomonitoring-and-gravity-field/projects/abgeschlossene-projekte/geo-research-satellite-gfz-1/

[7] Paolozzi A, Ciufolini I. LARES successfully launched in orbit: Satellite and mission description. Acta Astronautica. 2013;**91**:313-321. Available from: http://arxiv.org/ftp/arxiv/papers/1305/1305.6823.pdf

[8] Ciufolini I, Paolozzi A, König R, Pavlis EC, Ries JC, Matzner R, et al. Fundamental physics and general relativity with the LARES and LAGEOS satellites. Nuclear Physics B - Proceedings Supplements. 2013;**243-244**:180-193

[9] Burmistrov VB, Vasiliev VP, Parkhomenko NN, Shargorodsk VD. REFLECTOR, LARETS and METEOR-3M missions. In: IPIE Presentation at 14th International Workshop on Laser Ranging. San Fernando, Spain; 2004. Available from: https://cddis.nasa.gov/lw14/docs/presnts/tar3a_vbp.pdf

[10] CHAMP Mission Website. Available from: https://www.gfz-potsdam.de/en/section/geomagnetism/infrastructure/champ/satellite-systems/

[11] Tapley B. GRACE—Gravity Recovery and Climate Experiment. NASA. 2011. Available from: http://solarsystem.nasa.gov/scitech/display.cfm?ST_ID=186

[12] Tapley BD, Bettadpur S, Watkins M, Reigber C. The gravity recovery and climate experiment: mission overview and early results. Geophysical Research Letters. 2004;**31**

[13] Webb F, Flechtner F, Landerer F, Morton P, Watkins M, Save H, et al. Gravity recovery and climate experiment follow-on mission. In: Proceedings of the 70th IAC (International Astronautical Congress); Washington DC, USA; 21-25 October 2019. Paper: IAC-19.B1.1. Available from: https://iafastro.directory/iac/proceedings/IAC-19/IAC-19/B1/2/manuscripts/IAC-19,B1,2,4,x54184.pdf

[14] Unwin M. TechDemoSat-1 and the GNSS reflectometry experiment. In: TechDemoSat-1 User Consultation Meeting. Southampton, United Kingdom: National Oceanography Centre; 2015. Available from: http://www.merrbys.co.uk:8080/CatalogueData/Documents/TDS1%20SGRReSI%20Experiment.pdf

[15] Rose R, Wells W, Rose D, Ruf C, Ridley A, Nave K. Nanosat technology and managed risk: An update of the CYGNSS microsatellite constellation mission development. In: Proceedings of the AIAA/USU Conference on Small Satellites. Logan, Utah, USA; 2-7 August 2014. Paper: SSC14-VI-4. Available from: http://digitalcommons.usu.edu/cgi/viewcontent.cgi?article=3103&context=smallsat

[16] Ruf C. Two Year Anniversary of CYGNSS on Orbit. NASA Earth Observatory. 2018. Available from: https://earthobservatory.nasa.gov/blogs/fromthefield/2018/12/15/two-year-anniversary-of-cygnss-on-orbit/

[17] Ruf C, Asharaf S, Balasubramaniam R, Gleason S, Lang T, McKague D, et al. In-orbit performance of the constellation of CYGNSS hurricane satellites. Bulletin of the American Meteorological Society. 2019. DOI: 10.1175/BAMS-D-18-0337.1

[18] Barney RD et al. DRAFT science instruments, observatories, and sensor systems roadmap. In: Technology Area 08; NASA; 2008

[19] Touboul P, Rodrigues M, Metris G, Tatry B. MICROSCOPE, testing the equivalence principle in space. Comptes Rendus de l'Academie des Sciences, Series IV Physics. 2001;**2**(9)

[20] Conklin J, Nguyen A. Drag-free control and drag force recovery of small satellites. In: 31st AIAA Utah Small Satellite Conference. Utah; 2017

[21] Danzmann K, Rudiger A. LISA technology—Concept, status, prospects. Classical and Quantum Gravity. 2003;**20**

[22] Everitt CWF et al. Gravity probe B: Final results of a space experiment to test general relativity. Physical Review Letters. 2011;**106**(22)

[23] Canuto E. Drag-free and attitude control for the GOCE satellite. Automatica. 2008;**44**(7)

[24] Briess K, Kornemann G, Wickert J, MicroGEM: Microsatellites for GNSS earth monitoring. Abschlussbericht Phase 0/A; 2009. p. 244. Available from: https://www.gnss.tu-berlin.de/menue/forschung/aktuelle_projekte/gnss_fernerkundung_mit_kleinsatelliten/

[25] Høeg P, Fragner H, Dielacher A, Zangerl F, Koudelka O, Høeg P, et al. PRETTY: Grazing altimetry measurements based on the interferometric method. In: Proceedings of the 5th Workshop on Advanced RF Sensors and Remote Sensing Instruments (ARSI'17); European Space Agency, GPS Reflectometry; 2017. Available from: https://backend.orbit.dtu.dk/ws/portalfiles/portal/134709739/ARSI17_Fragner_PRETTY_2017.pdf

[26] Pavlis EC, Merkowitz SM, Beaudoin CJ, Kuzmicz-Cieslak M, Rowlands DD, Lemoine FG. GEOCON: Geodetic system ties using a CubeSat constellation. Geophysical Research. 2019;**21**:EGU2019-6158-1. EGU General Assembly 2019. Available from: https://meetingorganizer.copernicus.org/EGU2019/EGU2019-6158-1.pdf

[27] Conklin J, Nguyen A. Drag-free control and drag force recovery of small satellites. In: 31st AIAA Utah Small Satellite Conference. Utah; 2017. Available from: https://digitalcommons.usu.edu/cgi/viewcontent.cgi?article=3691&context=smallsat

[28] Zanoni C, Alfauwaz A, Aljadaan A, Althubiti S, Balakrishnan K, Buchman S, et al. The design of a drag-free CubeSat and the housing for its gravitational reference sensor. In: 2nd IAA Conference on University Satellites Missions and CubeSat Workshop. 2016. IAA-CU-13-12-01. Available from: https://www.researchgate.net/publication/303331084_The_Design_of_a_Drag-Free_CubeSat_and_the_Housing_for_its_Gravitational_Reference_Sensor

Chapter 9

Nanosatellites and Applications to Commercial and Scientific Missions

Adriano Camps

Abstract

In the past two decades, a silent revolution has taken place in the space domain, leading to what today is known as "New Space." We have passed from a selected group of countries, space agencies, and big industries building, launching, and operating satellites and other spacecrafts, of a scenario in which many universities and research institutes can do it. The key of this was the definition of the "CubeSat" standard, back to 1999. In 2013, it all took off on the commercial Earth Observation sector with the first launches from two companies that are now running 100+ CubeSat constellations for optical imaging or weather prediction, with very low revisit times. Today, the same revolution is taking place in the fields of Telecommunications, and Astronomical Scientific missions. In this chapter, the evolution of the space sector is briefly revised until the arrival of the CubeSats. Then, the CubeSat intrinsic limitations are discussed as they are key to understand the development and current situation of the CubeSat sector. NASA and ESA strategies are also presented. The chapter concludes with a summary of the technology roadmap to enable the next generation of CubeSat-based missions, including satellite constellations or federations, formation flying, synthetic apertures...

Keywords: satellites, CubeSats, mission, earth observation, astronomy, planetary exploration, enabling technologies

1. Introduction

1.1 From the sputnik to the CubeSats

At the beginning of the space age, all satellites were "small." Sputnik 1 was the first artificial Earth satellite (**Figure 1a**) [1]. It was launched by the Soviet Union from Baikonur Cosmodrome on October 4, 1957, into an elliptical low Earth orbit (LEO) with an inclination of 65°. Sputnik 1 was a 58-cm-diameter metal sphere, weighing approximately 84 kg, with four radio antennas transmitting at 20.005 and 40.002 MHz. Tracking and studying Sputnik 1 signals from Earth provided valuable information on upper atmosphere density, and the propagation of radio signals provided information on the ionosphere. Sputnik did not have solar panels, so the mission ended after 3 weeks when batteries died.

IntechOpen

Explorer 1 was the first US satellite (**Figure 1b**) [2], and the third one after Sputnik 1 and 2. It was launched from Cape Canaveral, Florida, on January 31, 1958. Explorer 1 was 205 cm tall and 15 cm in diameter, weighing nearly 14 kg. It was the first spacecraft to detect the Van Allen radiation belts. Explorer 1 did not have solar panels either, so after 4 months the mission ended when batteries were exhausted.

Vanguard 1 was the fourth artificial Earth satellite (**Figure 1c**) [3]. It was launched by the USA from Cape Canaveral on March 17, 1958, into a 654 by 3969 km elliptical orbit with an inclination of 34.25°. Vanguard 1 was a 16.5-cm-diameter aluminum sphere, weighing just 1.47 kg, and it was the first satellite with six solar cells powering two beacons at 108 and 108.03 MHz, which were used to measure the total electron content.

During the first two decades of the space age, each satellite had its own design. They were the art pieces of the space craftsmen. Standard spacecraft busses were practically unknown until the end of the 1970s. In the early 1980s, microsatellites emerged and adopted a radically different design approach to reduce costs, focusing on available and existing technologies and using properly qualified commercial off-the-shelf (COTS) components.

For many years, satellite mass increased as illustrated in **Table 1**. However, except for some military, astronomy, and specific communication applications, it seems that the era of massive satellites is over.

The "small satellite mission philosophy" represents a design-to-cost approach, with strict cost and schedule constraints, often combined with a single mission objective in order to reduce complexity. **Figure 2** from [14] summarizes the standardized definition of satellites according to their weight: picosatellites (0.1–1 kg), nanosatellites (1–10 kg), microsatellites (10–100 kg), and mini-satellites or small/medium satellites (100–1000 kg).

In the field of Earth observation (EO), this has led to smaller target-focused missions which, with reduced spacecraft and launch costs (shared rides), are enabling massive (>100) satellite constellations of nano- and microsatellites with reduced revisit times, unthinkable just a few years ago.

In the field of satellite communications, there are plans as well to deploy massive constellations of LEO satellites to provide worldwide Internet coverage, IoT services, and machine-to-machine (M2M) communications.

It is anticipated that enhanced inter-satellite communication capabilities (LEO-ground, LEO-LEO, LEO-MEO, and LEO-GEO) will also improve the performance of EO systems [15]. All this is leading to the evolution of the space segment from monolithic to distributed and federated satellite systems [16], aiming at establishing win-win collaborations between satellites to improve their mission performance by using the unused onboard resources.

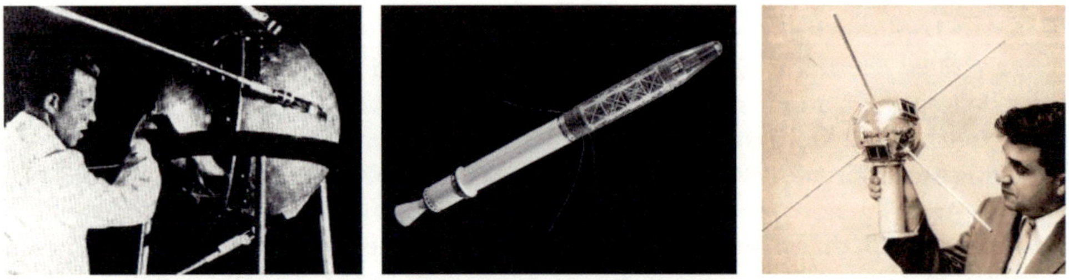

a) b) c)

Figure 1.
Pictures of (a) sputnik 1 [4], (b) explorer 1 [5], and (c) vanguard 1 [6].

Spacecraft	Agency application	Mass	Duration
KH-11 Kennen (a.k.a. CRYSTAL, EECS, 1010) [7]	US NRO/optical imaging	19,600 kg	1976–present
Proton [8]	USSR/astronomy	17,000 kg	1965–1969
Compton Gamma Ray Obs. [9]	US NASA/astronomy	16,329 kg	1991–2000
Lacrosse [10]	US NRO/SAR	14,500–16,000 kg	1988–2005
Hubble Space Telescope [11]	US NASA/astronomy	11,110 kg	1990–present
ENVISAT [12]	ESA/Earth observation	8211 kg	2002–2012
Telstar 19 V [13]	Canada/communications	7075 kg	2018–present

Table 1.
Heaviest spacecrafts (excluding space stations and manned orbiters).

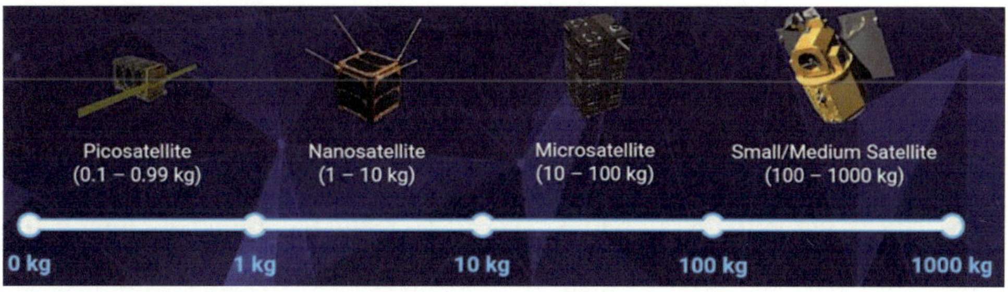

Figure 2.
Satellite classification [14].

1.2 The CubeSat standard

The so-called CubeSat standard was conceived in 1999 by Profs. Jordi Puig-Suari of California Polytechnic State University (CalPoly) and Bob Twiggs of Stanford University to allow graduate students to conceive, design, implement, test, and operate in space a complete spacecraft in a "reasonable" amount of time (i.e., the duration of their studies). CubeSats are small satellite multiples of 1 U (1 U: 10 cm × 10 cm × 11.35 cm, weighing less than 1.33 kg), including all the basic subsystems as large satellites but using COTS components. The CubeSat "standard" only defines the mechanical external interfaces, i.e., those referring to the orbital deployer. Originally, it was never meant to be a standard, however, because of its simplicity, it soon became a "de facto" standard. As Prof. Twiggs said in an interview to Spaceflight Now in 2014: "*It all started as a university education program satellite. It was kind of funny. I didn't think that people would criticize it as much as they did, but we got a lot of feedback (...). Another thing that was kind of funny we had no interest from NASA or any of the military organizations. It just wasn't anything they were interested in, so it was all funded without any funding from those aerospace organizations.*" The first six CubeSats were launched on a Russian Eurockot on June 30th, 2003. Then, after more than a decade in which the concept silently matured in university labs, space agencies got interested and showed that CubeSat-based mission reliability could be improved by proper engineering. In 2013, it all took off on the commercial Earth Observation sector with the first launches from two companies that are now running 100+ CubeSats constellations for optical imaging or weather prediction, with very low revisit times. Today, many of the initial CubeSat limitations (most notably size, available power, and down-link bandwidth) are being overcome, and the same revolution is starting to take place in the fields of telecommunications, and astronomical scientific missions.

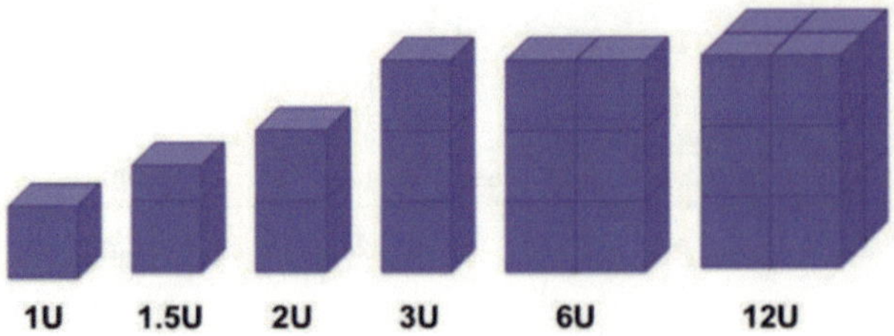

Figure 3.
CubeSat form factors from 1 U to 12 U [20].

The current CubeSat Design Specification defines the envelopes for 1 U, 1.5 U, 2 U, 3 U and 3 U+, and 6 U form factors (see, e.g., CubeSat Design Specification Rev. 13 or 6 U CubeSat Design Specification in [17], **Figure 3**), and the standardization of 12 U and 16 U is in progress, although some companies have produced standards up to 27 U [18]. On the other side, smaller picosatellites, the so-called PocketQubes, about 1/8 the size of a CubeSat, have also been standardized [19].

Probably, what has had the most significant impact in the popularization of the CubeSat standard has been the capability to separate the interface between the spacecraft and the poly-picosatellite orbital deployer (P-POD) and between the dispenser and the rocket itself. There are two different classes of PODs. The first type is the classical one with four rails in the corners [17], and the second one is with tables [18]. Note however that modern deployers from ISIS and NanoRacks allow larger deployables, wider solar panels, and thinner rails as compared to original P-POD, e.g., increased extruded height up to 9 mm and up to 2 kg per 1 U.

As of June 2019, 64 countries have launched nanosatellites or CubeSats. The total number of nanosatellites launched is 1186, from which 1088 are CubeSats. Most of them (273) have been launched from the International Space Station at ~400 km orbital height with an inclination of 51.6° and the rest at low Earth orbits (LEO) typically at 500 km sun-synchronous orbit (SSO) with an inclination of 97.5° (217 CubeSats) and 580 km height with 97.8° inclination (80 CubeSats). So far, only two (MarCO-1 and MarCO-2) have performed interplanetary missions.

1.3 Current status

Figure 4 shows the number of nanosatellites launched per year (a) and organization, either companies, universities, space agencies, etc., or (b) form factor from picosats, 0.25 up to 16 U.

As it can be appreciated, until 2013 most CubeSats were launched by universities and research institutes, and most of them were 1 U or 2 U. However, in 2013 the first 3 U CubeSats from the Planet Labs Inc. [22] and Spire Global Inc. [23] were launched. That was the beginning of today's revolution in EO, and—as of June 10, 2019—these two companies had launched the largest commercial constellations ever with 355 and 103 CubeSats, respectively. The following ones have launched at most seven CubeSats. Therefore, 3 U CubeSats are dominating the scene, and they will over the next decade, followed by far by the 1 U, 2 U, and 6 U form factors (**Figure 5**). However, it is expected that the next wave of growth will be based on 6 U and 12 U CubeSats, which offer the right balance between very capable payloads and limited manufacturing and launch costs.

Table 2 (extracted from the database in [21]) shows the main companies that have launched CubeSats, the number of launched and planned CubeSats, the year of the first launch, the form factor, the application field, and some technical details. The rows marked in light blue correspond to EO optical imaging, in light green to EO

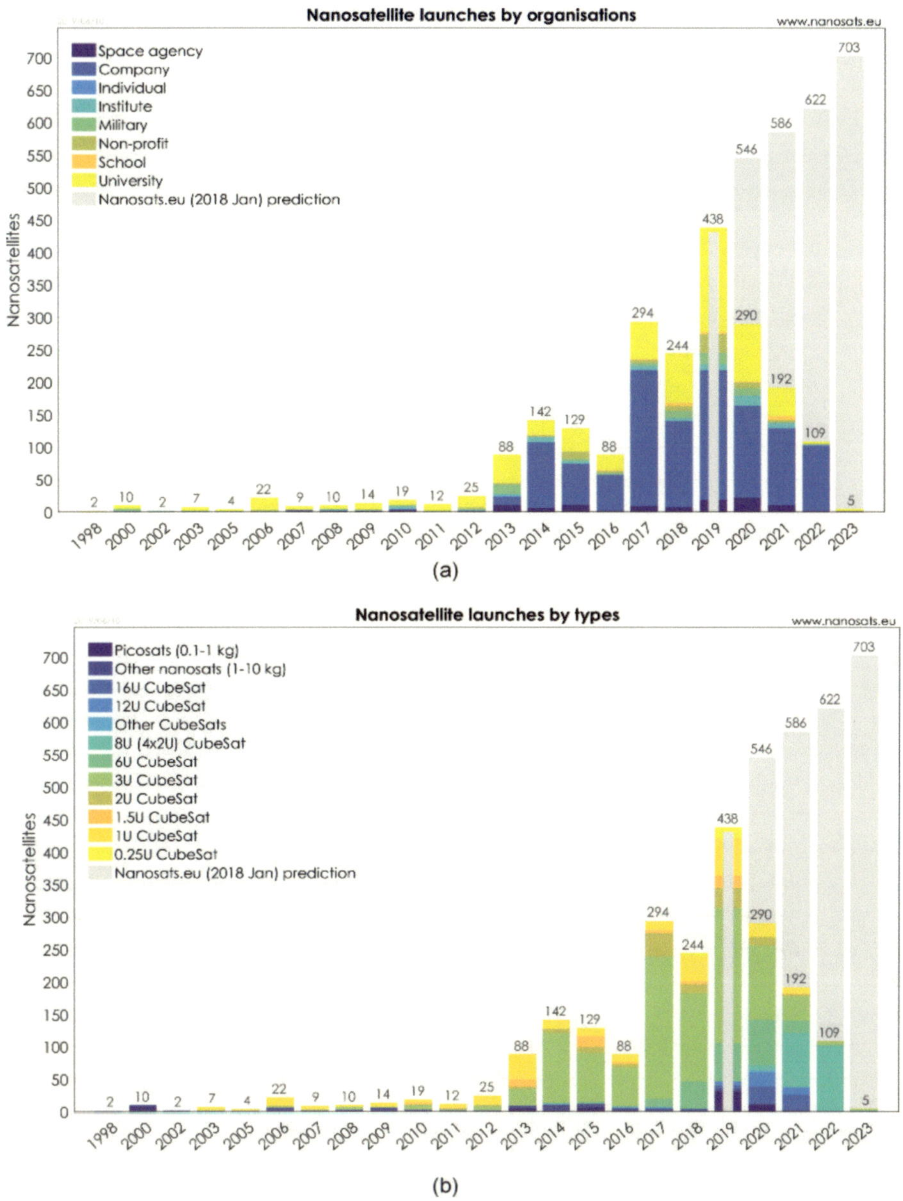

Figure 4.
The number of nanosatellites launched per year and (a) organization or (b) form factor [21].

passive microwaves applications, in dark green to EO active microwaves applications, and in light orange typically to IoT and M2M communications. In the next sections, we will focus on the EO applications but keeping in mind that future advances in satellite communication networks will also improve the performance of EO systems and enable new ones as well as distributed ones (e.g., large synthetic apertures).

The interested reader is encouraged to consult [21] for the most updated information as these numbers can change rapidly. Note that the number of CubeSats that can be launched in a single rocket can be very high. The current record is held by the Indian rocket PSLV-C37 that, on February 15, 2017, launched Cartosat-2D and 103 CubeSats, from which 88 are from the Planet Labs Inc. and 8 are from the Spire Global Inc. The interested reader is invited to see the deployment of these satellites from the onboard camera at [24].

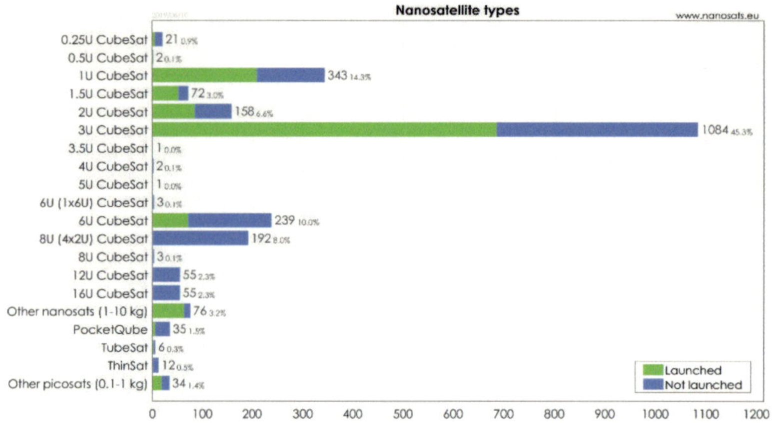

Figure 5.
The number of CubeSats by form factor [21].

Organization	Launched/ planned size	First launch	Form factor	Field	Technical and comments
Planet Labs	355/150	2013	3 U	Earth observation	29 MP sensor taking images with 3.7 m ground resolution and swath of 24.6 km × 16.4 km from 475 km altitude
Spire	103/150	2013	3 U	Weather, AIS, ADS-B, earthquake	Measure change in GPS signal after passing atmosphere to calculate precise profiles for temperature, pressure, and humidity. Investigating earthquake (ELF) detection
AprizeSat	12/12	2002	Microsat	IoT/M2M	Low-cost satellite data services for monitoring the fuel level and oil and gas pipelines and mobile tracking of shipping containers, railcars, and trailers
GeoOptics	7/N/A	2017	6 U	Weather	Using GPS radio occultation for weather data
Swarm Technologies	7/150	2018	0.25 U, 1 U	IoT/M2M	World's smallest two-way communication satellites
Commsat	7/72	2018	Microsat, 6 U, 3 U	IoT/M2M, AIS	Ladybeetle 1 is 100 kg and 3 CubeSats of 6 U and 3 of 3 U. Plans 4 more in 2019 and complete 72 satellites in 2022
Astro Digital	6/25	2014	6 U, 16 U	Earth observation	6 U has 22 m resolution in RGB and NIR. 16 U has 2.5 m resolution in RGB, red edge, and NIR with 70 MP sensor
Fleet Space	4/100	2018	3 U, 12 U, 1.5 U	IoT/M2M	Main constellation potentially with 12 U CubeSats
Sky and Space Global	3/200	2017	8 U, 6 U, 3 U	IoT/M2M	Communication service (voice, data, and M2M). Plans to use inter-satellite links

Organization	Launched/ planned size	First launch	Form factor	Field	Technical and comments
NanoAvionics	2/72	2017	6 U, 12 U	IoT/M2M	Global IoT constellation-as-a-service system aimed at IoT/M2M communication providers
Helios Wire	2/30	2017	6 U, 16 U	IoT/M2M	Uses 30 MHz of S-band spectrum to receive tiny data packages from billions of sensors
Kepler Communications	2/140	2018	3 U, 6 U	IoT/M2M, Internet	IoT/M2M data communication network. Monthly fee based on the data amount. Hope to achieve rates of 1–40 Mbps
Analytical Space	1/N/A	2018	6 U	IoT/M2M, orbital data relay, optical comms.	In-orbit relays receiving radio and downlink to ground with laser communication enabling more data downlink from satellites
Hiber	2/48	2018	6 U	IoT/M2M	Sends small packets of data (140 characters, accompanied by time stamp, identifier, and location)
Guodian Gaoke	2/38	2018	6 U	IoT/M2M	Reliable and economical satellite IoT services and industry solutions for our customers
Astrocast	2/80	2018	3 U	IoT/M2M	Targeting L-band. Inter-satellite links. NanoSpace propulsion. Further 80 satellites in orbit by 2022
AISTech	2/150	2018	2 U, 6 U	IoT/M2M, ADS-B, AIS, IR imaging	Two-way comms., thermal imaging to detect forest fires, aviation tracking (ADS-B)
ICEYE	2/18	2018	Microsat	SAR	21-launch agreement with Vector Space Systems. 10-platform agreement with York Space Systems
Harris Corp.	1/12	2018	6 U	Weather	Immediate access to 3D wind data sets from Harris-owned HyperCubes
SIRION	1/N/A	2018	CubeSat	IoT/M2M	IoT/M2M constellation. Partnered closely with Helios Wire. Sharing spectrum and satellites
Reaktor Space	1/36	2018	6 U, 2 U	Earth observation, hyperspectral	Hyperspectral constellation for smart agriculture with 100's of spectral bands and 20 m resolution
Myriota	1/50	2018	CubeSat	IoT/M2M	Run unique, patented software which provides reliable, direct-to-satellite Internet of Things (IoT) connectivity

Organization	Launched/ planned size	First launch	Form factor	Field	Technical and comments
LaserFleet	1/192	2018	CubeSat	Internet, optical comms.	Provide reliable 1 Gbps communication rates to aircraft at altitude. Higher effective data rate at a lower cost than the best-in-class Ku/Ka/V
ADASpace	1/192	2018	Microsat CubeSat	Earth observation	Establish a global, minute-level updated Earth image data network consisting of 192 satellites
Orbital Micro Systems	1/40	2019	3 U	Weather	Weather constellation utilizes microwave technology to capture temperature and moisture measurements, refreshed and delivered every 15 minutes
Lacuna Space	1/32	2019	3 U, 6 U	IoT/M2M	IoT/M2M constellation. Selected Open Cosmos to build 3 U demonstrator

Blue for constellations for optical EO, light green for passive microwave EO, dark green for active microwave EO, and orange for IoT and M2M communications.

Table 2.
The main existing and planned CubeSats and microsat commercial constellations.

2. Science opportunities

As illustrated in **Table 2**, by 2010 the maturity achieved by CubeSats and dispensers/launchers, on one side, and by some EO technologies (high-resolution multispectral imagery and GNSS-RO), on the other side, made possible that a number of companies developed applications based on commercial constellations. Today, thanks to an intense technology R&D, the situation is completely different.

The reasons for this have been threefold. On one side, due to their small size, it has been difficult to include deployable solar panels so as to increase the electrical power generated, and, on the other side, it has been difficult to include large antenna reflectors and to transmit enough RF power so as to have a satisfactory space-to-Earth link budget. The third reason was the poor pointing accuracy that now has significantly improved thanks to miniaturized star trackers and reaction wheels. So far, these reasons have kept active optical (LIDAR) and active microwave sensors (RADAR) away from CubeSats, although it has to be stated that synthetic aperture radars (SAR) have been recently boarded in microsatellite platforms successfully (ICEYE, **Table 2**).

For spaceborne EO applications, frequency bands are restricted to those in which the atmosphere exhibits a high transmissivity, that is, the microwave and millimeter-wave parts of the radio spectrum and the long-wave infrared (LWIR), near infrared (NIR), and visible (VIS) parts of the spectrum, as illustrated in **Figure 6**.

For astronomical observations, ground-based observations are also limited to Earth's atmospheric windows in the radio and optical parts of the spectrum (**Figure 6**). Therefore, to explore the remaining parts of the EM spectrum, space-based observatories are required.

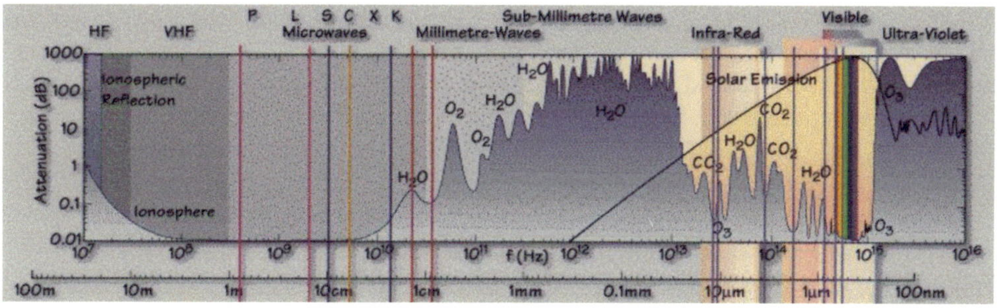

Figure 6.
Electromagnetic spectrum with different bands indicated [25].

In any case, either for EO or astronomical observations, the lower cost of individual CubeSat-based missions allows having more units, which reduces the revisit time at a given cost. This offers a number of new science opportunities [26]:

- Earth science:
  - Multipoint high temporal resolution of Earth processes
  - Mitigation of data gaps
  - Continuous monitoring
- Astrophysics:
  - Space telescopes allow access to energies across the whole electromagnetic spectrum avoiding large gaps in the radio, far IR, and the entire high-energy range (UV to γ-rays).
  - Feasibility to conduct time domain programs, which are very challenging with flagship missions such as the Hubble Space Telescope and James Webb Space Telescope.
  - Heliophysics, e.g., measurement of plasma processes in the magnetosphere-ionosphere system.
  - Planetary science: in situ investigation of planetary surfaces or atmospheres.
  - Astronomy and astrophysics: low-frequency radio science and the search for extrasolar planets.
- Biological and physical sciences, e.g., survival and adaptation of organisms to space

2.1 NASA science and technology strategy using CubeSats

Since the CubeSat standard was proposed in 1999, it took about a decade for NASA to start the Educational Launch of Nanosatellites (ELaNa) initiative in 2010. Partnerships were established with universities in the USA to design and launch CubeSats through NASA's CubeSat Launch Initiative (CSLI). Since then, 85 CubeSat missions have flown on 25 ELaNa calls, and 34 more CubeSats are manifested in 4 more calls. While it provides NASA with valuable opportunities to test emerging technologies that may be useful in future space missions, university students get involved in all phases of the mission, from the instrument and satellite design to its launch and monitoring.

As early as 2012, NASA's Science Mission Directorate (SMD) technology programs began to accommodate the use of CubeSats for validation of new science instruments and strategically promoted the use of small spacecraft to advance its science portfolio.

On one side, the Earth Science Technology Office (ESTO), which is responsible for identifying and developing technologies in support of future Earth Science Division missions, manages three major observation technology programs that solicit new awards on a 2–3-year selection cycle, as shown in **Table 3** [27].

And on the other side, following the outcomes of [28] in 2014, the Planetary Science Division (PSD) has also made significant strides toward accommodating small satellites for exploration of the solar system and for astrophysics research. **Table 4** [27] summarizes the three main planetary science technology programs.

The result of these continued investments is summarized in **Table 5**, where a number of EO techniques that were infeasible in 2012 [29] were all feasible 5 years later [30], many of them demonstrated by CubeSat missions, some of them commercial, and some even operational constellations. CubeSat-based astronomy missions will be discussed later.

Figure 7 illustrates some of these NASA CubeSat-based EO missions. They follow the 3 U or 6 U form factor and include deployable solar panels for higher electrical power generation capabilities. RainCube (**Figure 7c**) also includes a 0.5-m-diameter deployable Ka band that stows in 1.5 U. This antenna has a gain of 42.6 dBi, and it was optimized for the radar frequency of 35.75 GHz. References are provided for more information on the cited missions.

Earth science program	Approx. funding	Description
Instrument Incubator (IIP)	$28 M/year	Nurtures the development and assessment of innovative remote sensing concepts in ground, aircraft, or engineering model demonstrations (early to mid-stage development)
Advanced Components (ACT)	$5 M/year	Enables the research, development, and demonstration of component- and subsystem-level technologies to reduce the risk, cost, size, mass, and development time of missions and infrastructure
In-Space Validation of Earth Science Technologies (InVEST)	$5 M/year	Advances the readiness of existing Earth science-related technology and reduces risks to future missions through space flight validation using CubeSats

Table 3.
Earth science technology programs relevant to small satellites [27].

Planetary science program	Approx. funding	Description
Planetary Instrument Concepts for the Advancement of Solar System Observations (PICASSO)	$4 M/year	Supports the development of spacecraft-based instrument components and systems that show promise for future planetary missions. The program supports early-stage technologies
Maturation of Instruments for Solar System Exploration (MatISSE)	$6 M/year	Supports the advanced development of spacecraft-based instruments that may be proposed for future planetary missions that are at the middle stages of technology readiness
Development and Advancement of Lunar Instruments (DALI)	$5 M/year	Supports the development of science instruments for small lunar landers and orbital assets that are at the middle stages of technology readiness

Table 4.
Planetary science technology programs relevant to small satellites [27].

Technology	2012 technology review by Selva and Krejci	2017 technology review by Freeman et al.	Justification
Atmospheric chemistry instruments	Problematic	Feasible	PICASSO, IR sounders
Atmospheric temperature and humidity sounders	Feasible	Feasible	—
Cloud profile and rain radars	Infeasible	Feasible	JPL RainCube demo
Earth radiation budget radiometers	Feasible	Feasible	SERB, RAVAN
Gravity instruments	Feasible	Feasible	No demo mission
Hi-res optical imagers	Infeasible	Feasible	Planet Labs.
Imaging microwave radars	Infeasible	Problematic	Ka-Band 12 U design
Imaging multispectral radiometers (Vis/IR)	Problematic	Feasible	AstroDigital
Imaging multispectral radiometers (μW)	Problematic	Feasible	TEMPEST
Lidars	Infeasible	Problematic	DIAL laser occultation
Lightning imagers	Feasible	Feasible	—
Magnetic field	Feasible	Feasible	InSPIRE
Multiple direction/ polarization radiometers	Problematic	Feasible	HARP Polarimeter
Ocean color instruments	Feasible	Feasible	SeaHawk
Precision orbit	Feasible	Feasible	CanX-4 and CanX-5
Radar altimeters	Infeasible	Feasible	Bistatic LEO-GEO/ MEO
Scatterometers	Infeasible	Feasible	CYGNSS (GNSS-R)

In red: commercial companies.

Table 5.
EO technologies for CubeSat-based missions [29, 30].

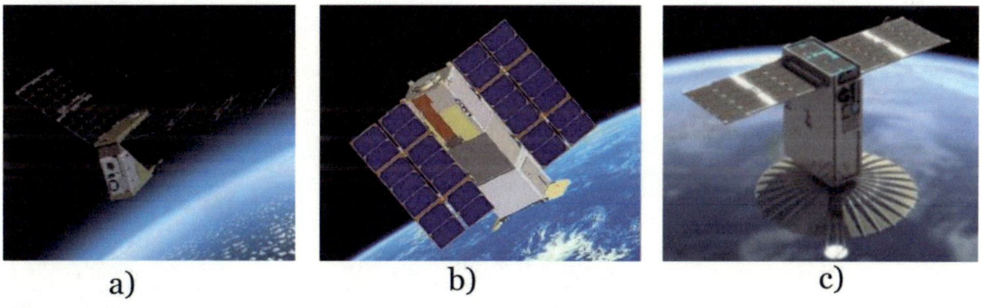

Figure 7.
Artist's view of (a) TEMPEST [31], (b) RAVAN [32], and (c) RainCube missions [33].

2.2 ESA science and technology strategy using CubeSats

On the educational side, the ESA launched in February 2008 the first Call for CubeSat Proposals to universities in ESA member and cooperating states. Seven student-built CubeSats were launched onboard the Vega maiden flight on February 13, 2012. Since then, 12 more CubeSats have been enrolled in the first and second editions of the "Fly Your Satellite!" program.

Since 2013, the ESA has invested more than 16 M€ as part of the General Support Technology Program (GSTP) FLY Element [34], in 12 CubeSat IOD missions [35, 36]. As part of ESA's Systems Department Project Office of the Systems

Department, Directorate of Technical and Engineering Quality, in April 2019, the CubeSat Systems Unit was created.

In addition to the work conducted by this unit, there are a number of other CubeSat-related initiatives in ESA:

- The Directorate of Telecommunications and Integrated Applications is developing a pioneer series of CubeSat missions, to test novel telecommunication technologies.
- The Directorate of Operations has OPS-SAT [37] ready to fly, an IOD test-bed for innovative mission control software.
- The Directorate of Human and Robotic Exploration is considering a CubeSat mission to test out a key capability for Mars sample return optical detection and navigation to a sample container from the orbit.
- The Science Directorate is also adapting some CubeSat technologies for operation in the deep space environment as well as studying the potential use of CubeSats in support of planetary science missions.
- The Directorate of Earth Observation will fly FSSCat [38, 39], a double 6 U CubeSat mission for tandem observation of the polar regions and for soil moisture mapping using the FMPL-3 (UPC, ES), a combined L-band microwave radiometer and GNSS-Reflectometer using a software-defined radio, and HyperScout-2 (Cosine, NL), a VNIR and TIR hyperspectral imager enhanced with artificial intelligence for cloud detection (PhiSat-1).

The first ESA CubeSat projects are listed in **Table 6**. In addition to these missions, numerous studies have focused on the applications of CubeSat missions and

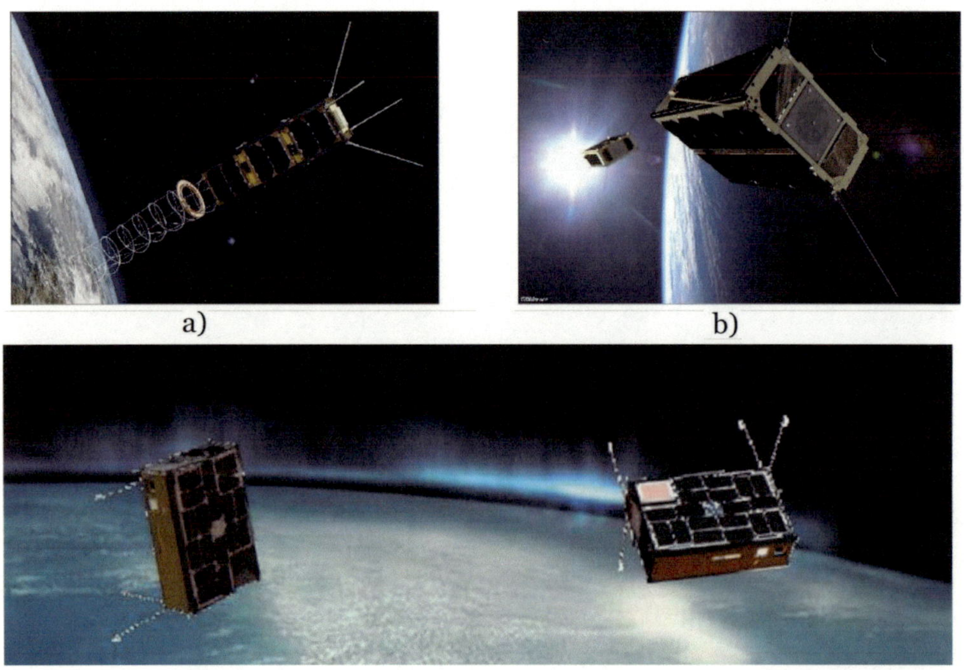

c)

Figure 8.
Artist's view of (a) GOMX-3 [35] and (b) GOMX-4 [36] nanosatellites in space (credits GomSpace) and (c) FSSCat mission [38, 39].

miniaturized payloads, including remote sensing with cooperative nanosatellites, asteroid impact missions, lunar CubeSats, astrobiology/astrochemistry experiment CubeSats, asteroid observer missions, etc.

Organization	Mission	Launch	Form factor	Field	Technical and comments
GomSpace (DK)	GOMX-3	2015	3 U	Tech demo	ADS-B, GEO Satcom signal monitoring, X-band transmitter (**Figure 8a**)
GomSpace (DK)	GOMX-4B	2018	2 × 6 U	Tech demo Earth observation	Inter-satellite link and propulsion while in tandem with GOMX-4A (GomSpace, Ministry of Defense, DK), star tracker HyperScout compact hyperspectral VNIR imager (Cosine, NL) (**Figure 8b**)
VKI (BE)	Qarman	2019	3 U	Tech demo	Demonstrates reentry technologies, novel heatshield materials, new passive aerodynamic drag stabilization system, and telemetry transmission during reentry via data relay satellites in low Earth orbit
RMI (BE) **KU Leuven (BE)**	SIMBA	2019	3 U	Earth observation	Total solar irradiance and Earth radiation budget
BIRA-IASB (BE) **VTT (FI)** **Clyde Space (UK)**	PICASSO	2019	3 U	Atmosphere and ionosphere	Stratospheric ozone distribution, mesospheric temperature profile, and ionospheric electron density
C3S and MTA EK (HU) **ICL (UK)** **Astronika (PO)**	RadCube	2019	3 U	Tech demo Space weather	3 U platform In situ space radiation and magnetic field in LEO
RUAG (AU) **TU Graz (AU)** **Seibersdorf Labor GmbH (AU)**	PRETTY	—	3 U	Earth observation	GNSS-R at low grazing angles, radiation dosimeter
ESA	OPS-SAT	2019	3 U	Tech demo	Experimentation with onboard and ground software by offering a safe and reconfigurable environment
UPC (ES) **Golbriak (EE)** **Deimos Eng (PT)** **Tyvak Intl. (IT)** **Cosine (NL)**	FSSCat	2019	2 × 6 U	Tech demo Earth observation	RF and O-ISL, federated satellite experiment ^{3}Cat-5/A: Microwave radiometer and GNSS-R (UPC, ES) ^{3}Cat-5/B: HyperScout-2 VNIR + TIR hyperspectral imager (Cosine, NL) (**Figure 8c**)

In blue from the CubeSat Systems Unit, Directorate of Technical and Engineering Quality; in orange from the Directorate of Operations; and in green from the Directorate of Earth Observation (2017 ESA S^3 Challenge, Copernicus Masters Competition).

Table 6.
The first ESA CubeSat-based missions.

3. Astronomy and interplanetary missions using CubeSats

As highlighted in Sections 1.3 and 2.1, the majority of the CubeSats orbiting today are devoted to Earth observation, notably from two commercial companies, followed by communications. In the coming years, these two categories will still dominate. Although the largest increase will occur in communication satellites, the growth in scientific (non-EO) missions will not be negligible (from 10 to 20%, **Figure 9**) considering that the predicted number of satellites to be launched is going to multiply by more than a factor of 3 (see **Figure 4**).

In particular, until 2017 there were only 5 astronomy missions, and in the field interplanetary missions, until 2018 only 14 nano−/microsatellites had been launched to destinations outside the LEO. Beyond-the-Earth orbit is the domain of civil agencies who, for the sake of reliability, have been historically reluctant to invest in small satellites. However, things may be changing, since only in 2018 four nano−/microsatellites made their way beyond the Earth orbit, which is more than those in the previous 5 years all together, and 35 more are expected to be launched in the coming 5 years. Naturally, most of them target the moon, but a non-negligible fraction will be devoted to interplanetary missions (**Figure 10**).

As in other fields, at the beginning all the astronomy or heliophysics missions were conducted by universities, and it was not until 2017 that the first NASA JPL mission (ASTERIA) was launched. Achieving state-of-the-art astronomy with CubeSats has become possible due to advances in precision pointing, communications technology, and deployables, among others (Tables 5.1 and 5.2 of [40]). **Table 7**, distilled from [21], shows the main astronomy and beyond-the-Earth past and planned missions. It also shows that the majority of these missions are based on the 6 U form factor, which is the smallest one capable to accommodate all the advanced attitude determination and control systems (ADCS), larger deployable solar panels and antennas, as well as telescope optics. It is also remarkable that so far there are no funded CubeSat missions in the far IR because the thermal stability and detector cooling require cryo-coolers for CubeSats that have yet to be developed for astrophysics due to power and space limitations [41].

It is worth noting that the large number of CubeSats to be launched to the Moon in 2020 corresponds to the Artemis-1 mission (**Figure 11**), formerly known as Exploration Mission-1. The first mission for NASA's Orion rocket and the European Service Module will send the spacecraft beyond the moon and back. Thirteen low-cost CubeSat missions were competitively selected as secondary payloads on the Artemis-1 test flight, all of them having the 6 U form factor. The selected CubeSats are Lunar Flashlight, Lunar South Pole, Near-Earth Asteroid Scout,

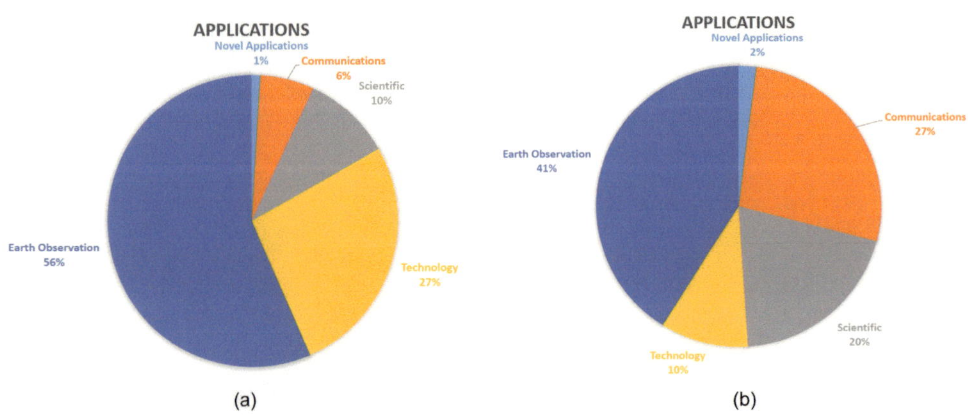

Figure 9.
Satellite application trends (1–50 kg): (a) 2014–2018 and (b) 2019–2023 (adapted from [14]).

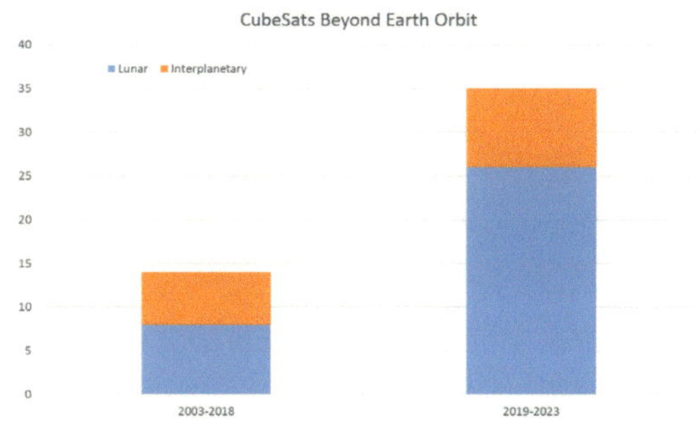

Figure 10.
CubeSats launched beyond the earth orbit: 14 from 2003 to 2018 and 35 planed from 2019 to 2023 (adapted from [14]).

Organization	Mission	Launch	Form factor	Technical and comments
Morehead State University Kentucky Space	CXBN	2012	2 U	• Cosmic X-ray background (CXRB) in the 30–50 keV range
University of Colorado	CSSWE	2012	3 U	• Measures the directional differential flux of solar energetic protons (SEPs) and Earth's radiation belt electrons
Austria Canada Poland	BRITE	2013 2014	8 U (2 × 2 × 2)	• BRITE Target Explorer Constellation: BRITE-Toronto, BRITE-Heweliusz, UniBRITE, BRITE-Austria, BRITE-Lem, BRITE-Montreal • Conducts photometric observations of some of the brightest stars in the sky to examine their variability. Observations will have a precision at least 10x better than achievable from ground-based observations
University of Colorado at Boulder	MinXSS	2015	3 U	• Provides spectral observations of the solar X-rays near the maximum of solar cycle 24 from 0.6 keV (20 Å) to 25 keV (0.5 Å)
JPL (USA) MIT (USA)	ASTERIA	2017	6 U	• Detects exoplanetary transits across bright stars • Pointing accuracy of ±0.003° (1σ) for 2 axes and ± 0.007° (1σ) for the third axis, with 0.5″ rms over 20 min, pointing repeatability of 0.001″ rms from orbit to orbit • ±0.01 K level temperature stability of the imaging detector
ERC, CNRS, ESEP Lab, PSL Université Paris, Fondation MERAC, CNES, CCERES and Obs. de Paris – LESIA	PicSat	2018	3 U	• To observe in visible light the potential transit of the • directly imaged giant planet β Pictoris b and perhaps even its moons and debris

Organization	Mission	Launch	Form factor	Technical and comments
University of Iowa	HaloSat	2018	6 U	• Measurement of soft X-ray emission from the hot halo of the Milky Way galaxy to resolve the missing baryon problem. Measure O VII and O VIII line emission in 400 fields (FOV ~ 10°) over 90% of the sky. Study of solar wind charge exchange emission to remove uncertainty on the oxygen line emission measurements • First NASA-funded astronomy mission
Spacety (China)	Tongchuan-1	2018	6 U	• Detects signals from gamma-ray bursts, to identify and locate the electromagnetic counterparts to gravitational waves
University of Colorado Boulder	MinXSS-2	2018	3 U	• As MinXSS
University of Colorado	CSIM	2018	6 U	• Observes the solar spectral irradiance
DARPA	SHFT-1	2018	3 U	• Collects radio-frequency signals in the HF (5–30 MHz) band to study the galactic background emissions, the HF signals from Jupiter, and the signals from terrestrial transmitters after having passed through the Earth's ionosphere
NASA	MarCO-1/ MarCO-2	2018	6 U	• Data relay to send data back to Earth during InSight's entry, descent and landing operations at Mars. Technology capability demonstration of communications relay system
University of Hawaii at Manoa	NEUTRON-1	2019*	3 U	• Measures low-energy neutron flux in LEO environment
Boston University	CuPID	2019*	6 U	• Miniaturized soft X-ray imaging telescope
University of Colorado Boulder	CUTE	2020	6 U	• To conduct a survey of exoplanet transit spectroscopy in the near UV of a dozen short-period, large planets orbiting F,G, and K stars to constrain stellar variability and measure mass loss rates • Second NASA-funded astronomy mission
Isaware (FI)	XFM Cube	2020	3 U	• Measuring X-ray fluxes
Lockheed Martin	LunIR	2020	6 U	• Lunar flyby to collect spectroscopy and thermography (MWIR sensor) for surface characterization, remote sensing, and site selection
Arizona State University	LunaH-Map	2020	6 U	• High-resolution mapping of hydrogen content of the entire south pole of the moon, including permanently shadowed regions up to a meter beneath the lunar surface

Organization	Mission	Launch	Form factor	Technical and comments
NASA JPL	Lunar Flashlight	2020	6 U	• Illuminates with lasers in four different bands the permanently shadowed regions and detect water ice absorption bands in the near-infrared
Morehead State University	Lunar IceCube	2020	6 U	• Prospects for water ice and other lunar volatiles as a function of time of day, latitude, and regolith composition/mineralogy from a low-perigee lunar orbit flying only 100 km (62 miles) above the lunar surface
Arizona State University (USA)	SPARCS	2021	6 U	• Monitoring in the far (153–171 nm) and near UV (258–308 nm) of low-mass stars (0.2–0.6 MΘ); the most dominant hosts of exoplanets • Each star observed for at least one stellar rotation (4–45 days) • Third NASA-funded astronomy mission
NASA's Goddard Space Flight Center	BurstCube	2021	6 U	• Detection of gamma ray transients in the 10–1000 keV energy range. Valuable capability to catch the predicted counterparts of gravitational wave sources • Fourth NASA-funded astronomy mission
ESA **Luxembourg Space Agency (LU)** **GomSpace (DK)**	M-ARGO	2023	12 U	• Demonstrating asteroid rendezvous and identifying in situ resources with multispectral imager and laser altimeter
ESA	HERA CUBESAT	N/A	2x6 U	• Observing asteroid and deflection assessment

Table 7.
Non-comprehensive list of astronomy and beyond-the-earth CubeSat-based missions.

BioSentinel (carrying the first living creatures into deep space since 1972), SkyFire, Lunar IceCube, CubeSat for Solar Particles (CuSP), Lunar Polar Hydrogen Mapper (LunaH-Map), EQUULEUS, OMOTENASHI, ArgoMoon, Cislunar Explorers, Earth Escape Explorer (CU-E^3), and Team Miles.

Talking about interplanetary missions, on May 5, 2018, NASA launched a stationary lander called InSight to Mars. InSight landed on Mars on November 26, 2018. Riding along with InSight were two CubeSats—the first of this kind of spacecraft ever to fly to deep space [42]. Both MarCO-A and MarCO-B succeeded in a flyby of Mars, relaying data to Earth from InSight as it landed on Mars. **Figure 12** shows an artist view of the MarCOs with the reflectarray used for communication purposes.

In addition to the "classical" astronomy, lunar and Martian missions cited above, CubeSats are nowadays finding their way to other bodies of the solar system, and there are proposals [43] to send them to Venus (CUVE mission), Deimos and Phobos asteroids (PRISM and PROME missions), comets (PrOVE mission), or Jupiter (ExCSITE mission, [44]). **Figure 13** from [44] illustrates the LEO and beyond-LEO CubeSat exploration initiatives.

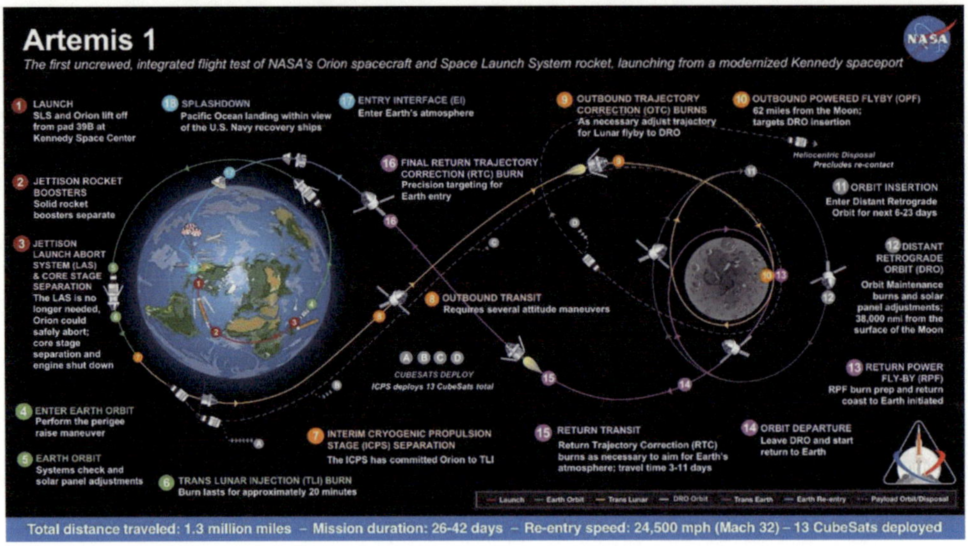

Figure 11.
Overview of the mission plan for Artemis-1: CubeSats will be deployed at steps A, B, C, and D [https://www.nasa.gov/image-feature/artemis-1-map].

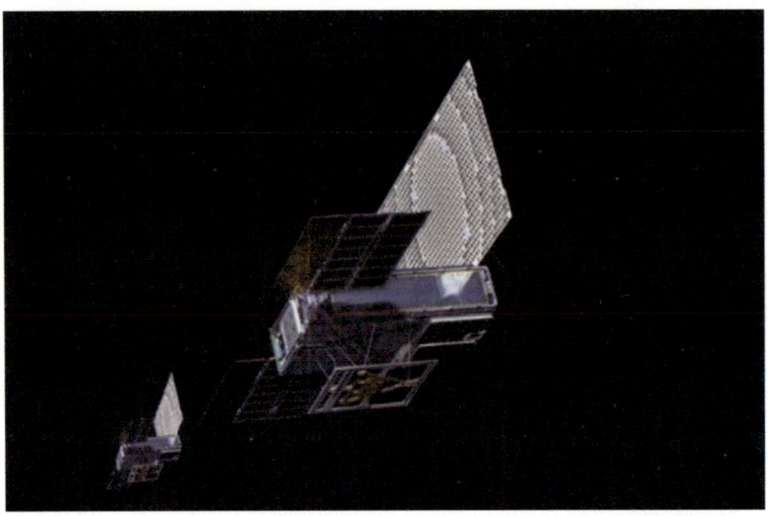

Figure 12.
Artist view of MarCO-A and MarCO-B [42].

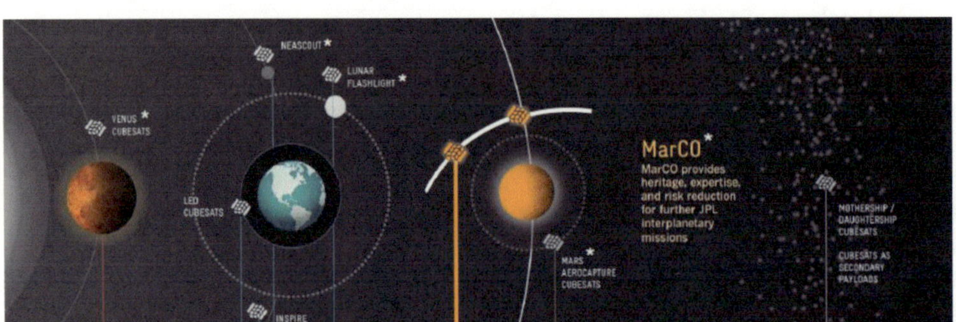

Figure 13.
Solar system exploration with CubeSats and nanosats [44].

4. Conclusions and way forward

Since its conception in 1999, CubeSats have produced a "disruptive innovation": from simple applications at the bottom of a market (mostly educational), they have relentlessly moved up, eventually displacing established medium-size competing satellites. However, CubeSats cannot displace all the large space missions as physics laws cannot be changed, i.e., large apertures and focal lengths are required to collect faint signals and achieve large angular resolution. However, CubeSats are finding their own niche in many Earth observation, astronomical, and communications applications where short revisit times or even continuous monitoring is required.

Early CubeSats typically had short lifetimes once in orbit (a few months), but with increased ground testing and added redundancies, lifetimes have grown significantly, up to 4–5 years in some cases.

Despite all these outstanding improvements, in order to exploit the full potential of CubeSats, many technologies still need to be developed. **Table 8** summarizes the enabling technologies required for different science applications, indicating in red the most challenging technologies and applications, notably increased communications performance, reliability, thermal stability, and calibration accuracy, to form constellations or formation flying satellite topologies to create large interferometers and distributed apertures.

Science discipline	Enabling technology	Example application
Solar and space physics	Propulsion	Constellation deployment and maintenance, formation flying
	Sub-arcsecond attitude control	High-resolution solar imaging
	Communications	Missions beyond low Earth orbit
	Miniature field and plasma sensors	In situ measurements of upper atmosphere plasmas
Earth science	Propulsion	Constellations for high-temporal resolution observations and orbit maintenance
	Miniaturized sensors	Stable, repeatable, and calibrated datasets
	Communications	High data rates
Planetary science	Propulsion	Orbit insertion
	Comms&Comms Infrastructure	Direct/indirect to Earth communications
	Radiation-tolerant electronics	Enhanced reliability in planetary magnetospheres, long flights
	Deployables	Deployable solar panel enhanced power generation
		Deployable mirrors and antennas
Astronomy and astrophysics	Propulsion	Constellations for interferometry, distributed apertures
	Sub-arcsecond attitude control	High-resolution imaging
	Communications	High data rate
	Deployables	Increased aperture and thermal control
	Miniaturized sensors	UV and X-ray imaging
Physical and biological	Thermal control	Stable payload environment

Table 8.
CubeSat-enabling technologies and potential applications for each science discipline (adapted from [40]).

In the field of Earth observation, future developments in nanosat sensors will likely occur:

- In the field of passive microwave sensors:

    - Miniature microwave and millimeter-wave radiometers for weather applications, such as the MiniRad which is onboard the Global Environmental Monitoring System (GEMS) constellation from Orbital Micro Systems [45], or

    - GNSS-R instruments with real-time processing for target detection/identification [46] or—as larger downlink bandwidths are available—with raw data acquisition and on-ground processing to optimize the processing according to the target, as planned in FMPL-3, the evolution of the FMPL-2 on board FSSCat [38, 39].

- In the field of passive VNIR/TIR hyperspectral imagers, imagers will include a larger number of bands but will include advanced image compression algorithms to minimize the amount of information to be downloaded and will incorporate artificial intelligence to download only the information extracted instead of the raw data.

Also, both their calibration will have to be refined so as to improve the quality of the scientific data.

Due to power and antenna size requirements, active microwave sensors (e.g., radar altimeters and SARs) will likely remain in domain of mini- and microsats (< 100 kg, e.g., ICEYE constellation [47]), and it is unlikely that active optical sensor technology (i.e., lidars) develops in small satellites in the midterm.

In the field of astronomy, and in particular heliophysics, NASA has also been taking the lead. In 2017 NASA selected nine proposals under its Heliophysics Small Explorers Program [48]: (1) the Mechanisms of Energetic Mass Ejection Explorer (MEME-X), (2) the Focusing Optics X-ray Solar Imager (FOXSI), (3) the Multi-Slit Solar Explorer (MUSE), (4) the Tandem Reconnection and Cusp Electrodynamics Reconnaissance Satellites (TRACERS), (5) the Polarimeter to Unify the Corona and Heliosphere (PUNCH), (6) the Atmospheric Waves Experiment (AWE), (7) the US Contributions to the THOR mission (THOR-US), (8) the Coronal Spectrographic Imager in the Extreme ultraviolet (COSIE), and (9) the Sun Radio Interferometer Space Experiment (SunRISE) mission concept, which is a space-based sparse array, composed of formation flying of six SmallSats designed to localize the radio emission associated with coronal mass ejections (CMEs) from the sun [49].

More recently, in August 2019, NASA selected two proposals to demonstrate SmallSat technologies to study interplanetary space [50]: (1) Science-Enabling Technologies for Heliophysics (SETH) will demonstrate two technologies, an optical communications technology and experiment to detect solar energetic neutral atoms as well as an array of waves and other particles that erupt from our sun, and (2) Solar Cruiser, which will deploy a nearly 18,000 square foot solar sail and a coronagraph instrument that would enable simultaneous measurements of the sun's magnetic field structure and velocity of coronal mass ejections or CMEs.

As a final thought, quoting Prof. Puig-Suari, "Before cubesats, we were so conservative nobody was willing to try anything out of the ordinary. When we did, we discovered some of the things everybody said would not work, did work. The fundamental change was that there was a mechanism to go try to those things. Some will work and some will not, but it allows us to try them and that was very infrequent before cubesats arrived. That was really important. That was the big change." And this is just the beginning of a new way to do Earth observation, astronomy, and satellite communications much more, in a different and more efficient way than it was done in the past decades. What will the future bring? Nobody knows, but certainly the future is being shaped today with these novel technologies, and only our imagination will set the limits.

Acknowledgements

This work has been supported by an ICREA Acadèmia award from the Generalitat de Catalunya to Prof. A. Camps. The author wants to express his gratitude to Profs. A. Golkar (Skoltech, Moscow, Russia) and H. Park (Universitat Politècnica de Catalunya, Barcelona, Spain) for the revision of this chapter and useful comments.

Author details

Adriano Camps[1,2]

1 CommSensLab "María de Maeztu" Unit, Department of Signal Theory and Communications, Universitat Politècnica de Catalunya UPC-Barcelona Tech, Barcelona, Spain

2 Institut d'Estudis Espacials de Catalunya, IEEC/CTE-UPC, Barcelona, Spain

*Address all correspondence to: camps@tsc.upc.edu

IntechOpen

© 2019 The Author(s). Licensee IntechOpen. This chapter is distributed under the terms of the Creative Commons Attribution License (http://creativecommons.org/licenses/by/3.0), which permits unrestricted use, distribution, and reproduction in any medium, provided the original work is properly cited.

References

[1] Sputnik-1 NSSDCA/COSPAR ID 1957-001B. Available from: https://nssdc.gsfc.nasa.gov/nmc/spacecraft/display.action?id=1957-001B [Accessed: 31 August 2019]

[2] Explorer-1 NSSDCA/COSPAR ID: 1958-001A. Available from: https://nssdc.gsfc.nasa.gov/nmc/spacecraft/display.action?id=1958-001A [Accessed: 31 August 2019]

[3] Vanguard-1 NSSDCA/COSPAR ID: 1958-002B. Available from: https://nssdc.gsfc.nasa.gov/nmc/spacecraft/display.action?id=1958-002B [Accessed: 31 August 2019]

[4] What was Sputnik One?. Available from: https://www.universetoday.com/78110/sputnik-one/ [Accessed: 31 August 2019]

[5] Explorer/Early Satellites. Available from: https://www.nasa.gov/mission_pages/explorer/explorer.html [Accessed: 31 August 2019]

[6] The first satellite to have solar electric power and the oldest man made object still in orbit. Available from: https://twitter.com/delgeo1/status/975032039840903168 [Accessed: 31 August 2019]

[7] NRO. Develop Acquire Launch Operate (NRO Brochure). 22 March 2016. Archived from the original (PDF) on 26 January 2017. Retrieved 24 April 2016. Available from: https://web.archive.org/web/20170126092253/http://www.nro.gov/about/nro/NRObrochure.pdf [Accessed: 31 August 2019]

[8] Proton-1 NSSDCA/COSPAR ID: 1965-054A. Available from: https://nssdc.gsfc.nasa.gov/nmc/spacecraft/display.action?id=1965-054A [Accessed: 31 August 2019]

[9] Compton Gamma-Ray Observatory NSSDCA/COSPAR ID: 1991-027B. Available from: https://nssdc.gsfc.nasa.gov/nmc/spacecraft/display.action?id=1991-027B [Accessed: 31 August 2019]

[10] Lacrosse/Onyx: Radar Imaging Reconnaissance Satellite. Available from: https://www.globalsecurity.org/space/systems/lacrosse.htm [Accessed: 31 August 2019]

[11] Hubble Space Telescope (HST) NSSDCA/COSPAR ID: 1990-037B. Available from: https://nssdc.gsfc.nasa.gov/nmc/spacecraft/display.action?id=1990-037B [Accessed: 31 August 2019]

[12] ENVISAT Mission Summary. Available from: https://earth.esa.int/web/guest/missions/esa-eo-missions/envisat/mission-summary [Accessed: 31 August 2019]

[13] Telstar 19 Vantage. Available from: https://www.telesat.com/our-fleet/Telstar-19-VANTAGE [Accessed: 31 August 2019]

[14] Spaceworks Enterprises Inc. Nano/Microsatellite market forecast, 9th ed. 2019. Available from: https://www.spaceworks.aero/wp-content/uploads/Nano-Microsatellite-Market-Forecast-9th-Edition-2019.pdf [Accessed: 31 August 2019]

[15] Ruiz-De-Azúa JA, Camps A, Calveras Augé A. Benefits of using mobile ad-hoc network protocols in federated satellite systems for polar satellite missions. IEEE Access. 2018;**6**:56356-56367. DOI: 10.1109/ACCESS.2018.2871516

[16] Golkar A, Cruz IL. The federated satellite systems paradigm: Concept and business case evaluation. Acta Astronautica. 2015;**111**:230-248. DOI: 10.1016/j.actaastro.2015.02.009

[17] CubeSat Developer Resources. Available from: http://www.cubesat.org/resources/ [Accessed: 31 August 2019]

[18] Planetary Systems Corporation. "PAYLOAD SPECIFICATION FOR 3U, 6U, 12U AND 27U." Available from: http://www.planetarysystemscorp.com/web/wp-content/uploads/2016/08/2002367D-Payload-Spec-for-3U-6U-12U-27U.pdf [Accessed: 31 August 2019]

[19] Alba Orbital, TUDelft, and Gauss Team. The PocketQube Standard. Issue 1, 7th of June, 2018. Available from: https://dataverse.nl/api/access/datafile/11680 [Accessed: 31 August 2019]

[20] NASA. What are SmallSats and CubeSats? Available from: https://www.nasa.gov/content/what-are-smallsats-and-cubesats [Accessed: 31 August 2019]

[21] Nanosats database. Available from: www.nanosats.eu [Accessed: 31 August 2019]

[22] Planet Labs Inc. website. Available from: https://www.planet.com/ [Accessed: 31 August 2019]

[23] Spire Global Inc. website. Available from: https://www.spire.com/en [Accessed: 31 August 2019]

[24] ISRO PSLV-C37 onboard camera view of 104 satellites deployment. Available from: https://www.youtube.com/watch?v=c0BpjPUT5FE [Accessed: 31 August 2019]

[25] Electromagnetic spectrum with different bands indicated. Available from: http://www.altimetry.info/radar-altimetry-tutorial/how-altimetry-works/frequencies-used-their-impacts/ [Accessed: 31 August 2019]

[26] Alkalai L, Norton C, Freeman A. CubeSats and small satellites as a vehicle for space innovation and exploration of space beyond earth orbit. In: 4thIAA Conference on University Satellite Missions and CubeSat Workshop –Rome, December 4-7. 2018. Available from: https://www.gaussteam.com/wordpress/wp-content/uploads/2018/02/20171204_Leon-Alkalai-1.pdf [Accessed: 31 August 2019]

[27] Seablom MS, Tan FW, Norton CD, Baker CE, Millar PS, Mercer CR, et al. Advancing technology for NASA science with small spacecraft. In: Proceedings of the 2019 IEEE International Geoscience and Remote Sensing Symposium, Yokohama, Japan, July 28–August 2. 2019

[28] Keck Institute for Space Studies. Small Satellites: A Revolution in Space Studies. Pasadena, CA: California Institute of Technology; July 2014. Available from: http://kiss.caltech.edu/final_reports/SmallSat_final_report.pdf [Accessed: 31 August 2019]

[29] Selva D, Krejci D. A survey and assessment of the capabilities of Cubesats for earth observation. Acta Astronautica. 2012;**74**:50-58

[30] Freeman T. Deep Space cubesats and nanosats at JPL. May 2017. Available from: https://icubesat.files.wordpress.com/2017/05/2017-a-1-1-201705300833-deep-space.pdf [Accessed: 31 August 2019]

[31] TEMPEST-D (Temporal Experiment for Storms and Tropical Systems Technology - Demonstration). Available from: https://directory.eoportal.org/web/eoportal/satellite-missions/t/tempest-d [Accessed: 31 August 2019]

[32] RAVAN (Radiometer Assessment using Vertically Aligned Nanotubes) Pathfinder Mission. Available from: https://directory.eoportal.org/web/eoportal/satellite-missions/r/ravan [Accessed: 31 August 2019]

[33] RaInCube (Radar In a CubeSat). A precipitation profiling mission. Available from: https://directory.

eoportal.org/web/eoportal/satellite-missions/r/raincube [Accessed: 31 August 2019]

[34] ESA's 'Cubesat central' for smaller missions into space. Available from: http://www.spacedaily.com/reports/ESAs_Cubesat_central_for_smaller_missions_into_space_999.html [Accessed: 31 August 2019]

[35] GOMX-3 (GomSpace Express-3). Available from: https://directory.eoportal.org/web/eoportal/satellite-missions/g/gomx-3 [Accessed: 31 August 2019]

[36] GomX-4 (GomSpace Express-4) Mission. Available from: https://directory.eoportal.org/web/eoportal/satellite-missions/g/gomx-4 [Accessed: 31 August 2019]

[37] OPS-SAT (Operations nanoSatellite). Available from: https://directory.eoportal.org/web/eoportal/satellite-missions/o/ops-sat [Accessed: 31 August 2019]

[38] Camps A et al. FSSCat, the 2017 Copernicus masters "ESA sentinel small satellite challenge" winner: A federated polar and soil moisture tandem Mission based on 6U Cubesats. In: IGARSS 2018-2018 IEEE International Geoscience and Remote Sensing Symposium, Valencia. 2018. pp. 8285-8287. DOI: 10.1109/IGARSS.2018.8518405

[39] Camps A et al. FSSCat: A Cubesat-based tandem mission for earth observation of the polar regions. In: ESA Living Planet Symposium, 13-17 May 2019, Milano, Italy. 2019. Available from: https://lps19.esa.int/NikalWebsitePortal/living-planet-symposium-2019/lps19/Agenda/AgendaItemDetail?id=9da26b75-2380-43a9-a1c4-d9adb53b76a1 [Accessed: 31 August 2019]

[40] Achieving Science with CubeSats: Thinking inside the Box. National Academies; 2016. Available from: https://www.nap.edu/catalog/23503/achieving-sciencewith-cubesats-thinking-inside-the-box [Accessed: 31 August 2019]

[41] Shkolnik E. On the verge of an astronomy CubeSat revolution. Nature Astronomy. 2018;**2**:374-378 Available from: https://www.nature.com/articles/s41550-018-0438-8 [Accessed: 31 August 2019]

[42] Mars Cube One (MarCO). Mission Overview. Available from: https://www.jpl.nasa.gov/cubesat/missions/marco.php [Accessed: 31 August 2019]

[43] NASA Planetary Missions and Concepts. Available from: https://cubesats.gsfc.nasa.gov/missions.html [Accessed: 31 August 2019]

[44] Freeman A. Deep space NanoSats—Positioned for exponential growth. In: Proceedings of the 4S Symposium, Valletta, Malta, May 30–June 6. 2016. Available from: https://trs.jpl.nasa.gov/bitstream/handle/2014/46097/CL%2316-1703.pdf?sequence=1&isAllowed=y [Accessed: 31 August 2019]

[45] Orbital Micro Systems Achieves Earth Observation Breakthroughs. Available from: https://www.globenewswire.com/news-release/2019/08/19/1903620/0/en/Orbital-Micro-Systems-Achieves-Earth-Observation-Breakthroughs.html [Accessed: 31 August 2019]

[46] Di Simone A, Park H, Riccio D, Camps A. Sea target detection using spaceborne GNSS-R delay-Doppler maps: Theory and experimental proof of concept using TDS-1 data. IEEE Journal of Selected Topics in Applied Earth Observations and Remote Sensing. 2017;**10**(9). DOI: 10.1109/JSTARS.2017.270535

[47] Iceye SAR Satellite Constellation. Available at: https://www.iceye.com/. Last visited October 6, 2019

[48] NASA Selects Proposals to Study Sun, Space Environment. Available from: https://www.jpl.nasa.gov/news/news.php?feature=6912 [Accessed: 31 August 2019]

[49] Alibay F, Hegedus AM, Kasper JC, Lazio TJW, Neilsen T. SunRISE status: Concept development update. 2018 IEEE Aerospace Conference, Big Sky, MT. 2018. pp. 1-11. DOI: 10.1109/AERO.2018.8396371

[50] NASA selects proposals to demonstrate smallSat technologies to study interplanetary space. Available at: https://phys.org/news/2019-08-nasa-smallsat-technologies-interplanetary-space.html. Last visited: October 7, 2019